计算机系列教材

于立萍 楚旺 编著

Java项目案例开发入门
（微课视频版）

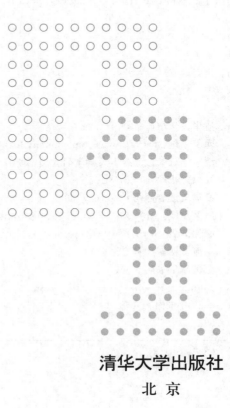

清华大学出版社
北 京

内 容 简 介

本书以实验案例和工程实训为主线,讲解 Java 面向对象技术及应用程序的开发,从软件工程的视角探索如何进行项目的设计及构建。全书共 15 章,主要内容包括:结构化程序设计、数组、类与对象、继承、多态、内部类、Java 常用类、枚举类、正则表达式、异常处理、集合、图形用户界面 Swing、JDBC 编程、输入/输出流、多线程及网络编程。本书最后以具体的工程应用项目为背景描述软件开发的详细流程,包括软件需求分析、体系结构设计、编码、测试及项目的发布。

本书突出与软件工程相结合的特点,既重视知识点的训练,又注重工程项目的实例分析与构建。本书不仅为读者提供了完整的代码,同时扫描书中的二维码可观看相关知识点的视频讲解。

本书可作为高等学校计算机专业及软件工程专业"Java 面向对象程序设计"课程的教材,也可以作为 Java 软件开发人员的参考书。

本书封面贴有清华大学出版社防伪标签,无标签者不得销售。
版权所有,侵权必究。举报:010-62782989,beiqinquan@tup.tsinghua.edu.cn。

图书在版编目(CIP)数据

Java 项目案例开发入门:微课视频版/于立萍,楚旺编著. —北京:清华大学出版社,2021.4
计算机系列教材
ISBN 978-7-302-57731-7

Ⅰ.①J… Ⅱ.①于… ②楚… Ⅲ.①JAVA 语言-程序设计 Ⅳ.①TP312.8

中国版本图书馆 CIP 数据核字(2021)第 050137 号

责任编辑:郭　赛
封面设计:常雪影
责任校对:焦丽丽
责任印制:宋　林

出版发行:清华大学出版社
网　　址:http://www.tup.com.cn,http://www.wqbook.com
地　　址:北京清华大学学研大厦 A 座　　　　邮　编:100084
社 总 机:010-62770175　　　　　　　　　　邮　购:010-83470235
投稿与读者服务:010-62776969,c-service@tup.tsinghua.edu.cn
质量反馈:010-62772015,zhiliang@tup.tsinghua.edu.cn
课件下载:http://www.tup.com.cn,010-83470236

印 装 者:三河市铭诚印务有限公司
经　　销:全国新华书店
开　　本:185mm×260mm　　印　张:21.75　　字　数:501 千字
版　　次:2021 年 6 月第 1 版　　　　　　　印　次:2021 年 6 月第 1 次印刷
定　　价:59.90 元

产品编号:084894-01

前 言

"新工科"建设已经成为当前工程教育改革的主流。信息科学,尤其是计算机科学、软件工程专业需要将工程教育的理念贯穿于教学环节,以培养学生系统设计及解决实际问题的工程能力。具体到程序设计语言类课程,需着眼于以实践能力为基础的"新工科"核心专业能力的培养。在此背景下,本书将面向对象编程思想与软件工程的过程实践融于一体,以实验案例和工程项目为载体,由浅入深,循序渐进,使读者在实践中理解和掌握 Java 面向对象技术,并提高解决软件工程实际问题的能力。

全书共 15 章,各章的主要内容如下。

第 1 章对 Java 语言进行概述,简单介绍 Java 平台、Java 语言特点及 Java 程序的开发环境。

第 2~4 章介绍 Java 语言的基本语法,包括基本的数据类型与运算、结构化程序设计语句及数组的定义与使用。

第 5、6 章是本书的重点内容,介绍 Java 面向对象的核心技术,主要包括类与对象、继承、接口与多态、内部类等。

第 7 章介绍枚举类与 Java 常用类,包括字符串、数学计算、随机数、日期等实用类。

第 8 章介绍正则表达式与异常处理知识。

第 9 章介绍 Java 的集合框架。由于集合框架中的接口与实现类众多,存储特性与性能也各不相同,所以本书在讲解的过程中特别注意集合类的区别,使读者能够根据程序的功能场景和性能需求选用合适的集合类。

第 10 章介绍 Java 的图形用户界面技术。

第 11 章介绍通过 JDBC 进行数据库编程的知识。通过 JDBC 可以很方便地操作各种主流数据库,利用 JDBC 编写的程序能够自动将 SQL 语句传送给相应的数据库管理系统,实现数据库的操作。

第 12 章介绍输入/输出流,重点突出流的基本概念,Java 通过数据流的方式实现不同数据源的统一操作,大大简化了数据的输入和输出操作。

第 13 章引入大量案例,系统介绍多线程的相关概念以及如何通过多线程解决系统的并发与同步问题。

第 14 章以多线程下载工具的实现为切入点,重点介绍网络编程的相关知识。

第 15 章以房屋租赁系统的设计与实现为综合案例,描述软件开发的详细流程,包括软件需求分析、体系结构设计、编码、测试及项目的发布。

为便于教学,本书提供教学视频、源代码、教学课件、教学大纲、习题参考答案等配套资源,读者可从清华大学出版社官方网站(www.tup.tsinghua.edu.cn)下载。

由于作者水平有限,书中难免有欠妥之处,敬请广大读者批评指正。

于立萍

2021 年 5 月

目 录

第1章 Java面向对象开发环境——搭建Java开发环境 1
  1.1 第一个Java应用程序——Hello World 1
  1.2 关键技术 1
    1.2.1 Java语言简介 1
    1.2.2 Java平台 1
    1.2.3 Java语言的特点 2
    1.2.4 Java小程序和应用程序 4
  1.3 搭建Java开发环境的基本步骤 4
    1.3.1 JDK安装与配置 4
    1.3.2 深入理解"HelloWorld"应用程序的执行 8
    1.3.3 Eclipse软件包的下载与安装 9
  1.4 练一练 10

第2章 基本数据类型及运算——身体质量指数计算器 11
  2.1 身体质量指数计算器 11
  2.2 程序设计思路 12
  2.3 关键技术 12
    2.3.1 Java标识符 12
    2.3.2 关键字 13
    2.3.3 Java的数据类型 13
    2.3.4 基本数据类型的转换 15
    2.3.5 运算符与表达式 16
    2.3.6 简单数据的输入与输出 20
  2.4 身体质量指数计算器设计步骤 21
  2.5 练一练 22

第3章 结构化程序设计——设计图书ISBN校验器 23
  3.1 图书ISBN校验器 23
  3.2 程序设计思路 23
  3.3 关键技术 24
    3.3.1 顺序结构 24

3.3.2　分支结构 ………………………………………………………………… 24
　　　3.3.3　循环结构 ………………………………………………………………… 28
　　　3.3.4　循环控制结构 …………………………………………………………… 30
　3.4　图书 ISBN 校验器设计步骤 …………………………………………………… 31
　3.5　练一练 …………………………………………………………………………… 33

第 4 章　数组——迷宫游戏 …………………………………………………………… 35
　4.1　迷宫游戏介绍 …………………………………………………………………… 35
　4.2　程序设计思路 …………………………………………………………………… 35
　4.3　关键技术 ………………………………………………………………………… 36
　　　4.3.1　什么是数组 ……………………………………………………………… 36
　　　4.3.2　数组的定义及初始化 …………………………………………………… 36
　　　4.3.3　数组的使用 ……………………………………………………………… 37
　　　4.3.4　数组的遍历 ……………………………………………………………… 38
　　　4.3.5　在方法中使用数组 ……………………………………………………… 39
　　　4.3.6　多维数组 ………………………………………………………………… 40
　　　4.3.7　Arrays 类 ………………………………………………………………… 41
　4.4　迷宫游戏设计步骤 ……………………………………………………………… 42
　4.5　练一练 …………………………………………………………………………… 45

第 5 章　类与类的继承——个人通讯录（一）……………………………………… 48
　5.1　个人通讯录（一）……………………………………………………………… 48
　5.2　程序设计思路 …………………………………………………………………… 48
　5.3　关键技术 ………………………………………………………………………… 48
　　　5.3.1　面向对象的基本概念 …………………………………………………… 48
　　　5.3.2　类的定义 ………………………………………………………………… 49
　　　5.3.3　对象 ……………………………………………………………………… 51
　　　5.3.4　方法的重载 ……………………………………………………………… 54
　　　5.3.5　构造方法 ………………………………………………………………… 54
　　　5.3.6　this 关键字 ……………………………………………………………… 56
　　　5.3.7　static 关键字 …………………………………………………………… 57
　　　5.3.8　初始化块 ………………………………………………………………… 59
　　　5.3.9　访问控制符 ……………………………………………………………… 60
　　　5.3.10　包的定义及其导入 ……………………………………………………… 61
　　　5.3.11　类的继承 ………………………………………………………………… 62
　　　5.3.12　子类重写父类方法 ……………………………………………………… 63
　　　5.3.13　super 关键字 …………………………………………………………… 64
　　　5.3.14　final 关键字 …………………………………………………………… 64

  5.3.15 继承与组合 65
 5.4 个人通讯录（一）设计步骤 66
  5.4.1 个人通讯录系统类图 66
  5.4.2 定义类 Contract 67
  5.4.3 定义类 Family 68
  5.4.4 定义类 Partner 69
 5.5 练一练 70

第 6 章　多态与内部类——个人通讯录（二） 72

 6.1 个人通讯录（二） 72
 6.2 程序设计思路 72
 6.3 关键技术 72
  6.3.1 多态 72
  6.3.2 instanceof 运算符 75
  6.3.3 抽象类与抽象方法 76
  6.3.4 接口 77
  6.3.5 内部类 81
  6.3.6 匿名内部类 82
 6.4 个人通讯录（二）设计步骤 83
  6.4.1 系统类图 83
  6.4.2 重构类 Contract 84
  6.4.3 重构类 Family 85
  6.4.4 重构类 Partner 86
  6.4.5 重构类 Company 87
  6.4.6 定义通讯录 PhoneBook 类 87
 6.5 练一练 90

第 7 章　Java 常用类与枚举类——21 点游戏 93

 7.1 21 点游戏介绍 93
 7.2 程序设计思路 93
 7.3 关键技术 94
  7.3.1 Object 类 94
  7.3.2 String 类 94
  7.3.3 StringBuilder 类与 StringBuffer 类 96
  7.3.4 Math 类 97
  7.3.5 Random 类 98
  7.3.6 Date 类 98
  7.3.7 Calendar 类 99

	7.3.8	SimpleDateFormat 类	100
	7.3.9	枚举类	100
7.4	21 点游戏设计步骤		102
	7.4.1	纸牌类	102
	7.4.2	玩家类	104
	7.4.3	游戏类	106
7.5	练一练		108

第 8 章　正则表达式与异常处理——个人通讯录（三）　111

8.1	个人通讯录（三）		111
8.2	程序设计思路		111
8.3	关键技术		111
	8.3.1	正则表达式	111
	8.3.2	异常概述	114
	8.3.3	异常处理	115
	8.3.4	自定义异常类	119
8.4	个人通讯录（三）设计步骤		120
	8.4.1	自定义异常子类	120
	8.4.2	Contract 类	120
8.5	练一练		122

第 9 章　集合——个人通讯录（四）　123

9.1	个人通讯录（四）		123
9.2	程序设计思路		123
9.3	关键技术		123
	9.3.1	集合概述	123
	9.3.2	Collection 接口	124
	9.3.3	集合的遍历	126
	9.3.4	Set 接口及其实现类	127
	9.3.5	List 接口及其实现类	130
	9.3.6	Map 接口及其实现类	131
	9.3.7	泛型	132
	9.3.8	Collections 工具类	133
9.4	个人通讯录（四）设计步骤		136
	9.4.1	重构联系人 Contract 类	136
	9.4.2	重构 Family 类	138
	9.4.3	重构 Partner 类	138
	9.4.4	重构通讯录 PhoneBook 类	138

9.5　练一练 ··· 140

第 10 章　基于 Swing 的图形用户界面——个人通讯录（五）　142
10.1　个人通讯录（五） ··· 142
10.2　程序设计思路 ··· 142
10.3　关键技术 ··· 143
　　10.3.1　图形用户界面与 Swing 概述 ·· 143
　　10.3.2　容器 ·· 144
　　10.3.3　常用组件 ··· 148
　　10.3.4　布局管理器 ·· 151
　　10.3.5　事件处理 ··· 155
　　10.3.6　菜单组件 ··· 160
　　10.3.7　表格组件 ··· 165
10.4　个人通讯录（五）设计步骤 ··· 170
　　10.4.1　项目总体结构 ·· 170
　　10.4.2　通讯录系统主界面 ··· 170
　　10.4.3　"编辑（添加）联系人"对话框 ··· 174
　　10.4.4　通讯录业务逻辑类 PhoneBook ··· 181
10.5　练一练 ·· 183

第 11 章　JDBC 编程——个人通讯录（六）　184
11.1　个人通讯录（六） ··· 184
11.2　程序设计思路 ··· 184
11.3　关键技术 ··· 184
　　11.3.1　JDBC 简介 ··· 184
　　11.3.2　MySQL 的安装 ··· 185
　　11.3.3　JDBC 常用接口及类 ·· 185
　　11.3.4　JDBC 操作数据库的基本步骤 ·· 186
11.4　个人通讯录（六）设计步骤 ··· 189
　　11.4.1　数据库设计 ·· 189
　　11.4.2　导入 MySQL 数据库驱动 ··· 191
　　11.4.3　项目总体结构 ·· 192
　　11.4.4　数据实体 bean ··· 192
　　11.4.5　工具包 util ·· 193
　　11.4.6　数据访问包 dao ·· 194
　　11.4.7　业务逻辑层 service ··· 201
　　11.4.8　图形用户界面层 gui ·· 205
11.5　练一练 ·· 205

第12章 输入/输出流——个人通讯录（七） 206

- 12.1 个人通讯录（七) 206
- 12.2 程序设计思路 206
- 12.3 关键技术 206
  - 12.3.1 File 类 206
  - 12.3.2 流的基本概念 207
  - 12.3.3 字节流与字符流 208
  - 12.3.4 包装流 209
  - 12.3.5 对象序列化及对象流 211
- 12.4 个人通讯录（七）设计步骤 213
  - 12.4.1 数据库设计 213
  - 12.4.2 数据实体类 Contract 213
  - 12.4.3 数据访问类 ContractDaoImpl 214
  - 12.4.4 图形用户界面 ContractDetailGUI 215
- 12.5 练一练 217

第13章 多线程——哲学家就餐问题 218

- 13.1 哲学家就餐问题介绍 218
- 13.2 程序设计思路 218
- 13.3 关键技术 218
  - 13.3.1 多线程的基本概念 218
  - 13.3.2 实现多线程的方法 219
  - 13.3.3 线程的生命周期 224
  - 13.3.4 线程的控制方法 226
  - 13.3.5 线程的同步 227
  - 13.3.6 线程通信 230
- 13.4 哲学家就餐问题程序设计步骤 231
  - 13.4.1 死锁的产生 231
  - 13.4.2 死锁解决方案 235
- 13.5 练一练 236

第14章 网络编程——多线程下载工具 237

- 14.1 多线程下载工具功能介绍 237
- 14.2 程序设计思路 238
- 14.3 关键技术 238
  - 14.3.1 HTTP 238
  - 14.3.2 URL 编程 239
  - 14.3.3 基于 TCP 的网络编程 241

  14.3.4 基于 UDP 的网络编程 ·············· 245
 14.4 多线程下载工具程序设计步骤 ················ 247
  14.4.1 下载任务实体类 ··············· 247
  14.4.2 主控界面 ······················ 248
  14.4.3 文件下载线程 ·················· 252
  14.4.4 文件下载子线程 ··············· 255
 14.5 练一练 ······································ 257

第 15 章 房屋租赁系统的设计与实现 ············ 258
 15.1 软件需求规格说明 ························ 258
  15.1.1 总体描述 ······················ 258
  15.1.2 具体需求 ······················ 259
 15.2 体系结构设计 ······························ 263
  15.2.1 Use Case 实现 ··············· 263
  15.2.2 逻辑视图 ······················ 264
  15.2.3 部署视图 ······················ 264
  15.2.4 实现视图 ······················ 264
  15.2.5 数据视图 ······················ 268
 15.3 编码实现 ···································· 270
  15.3.1 项目的文件结构 ·············· 270
  15.3.2 数据实体包 bean ············ 270
  15.3.3 数据访问包 dao ·············· 273
  15.3.4 业务逻辑包 service ········· 283
  15.3.5 工具包 util ··················· 290
  15.3.6 图形用户界面包 gui ········ 294
 15.4 测试 ·· 330
  15.4.1 搭建测试环境——导入 JUnit 包 ·········· 330
  15.4.2 单元测试 ······················ 331
 15.5 程序发布 ···································· 332
  15.5.1 打包项目 ······················ 332
  15.5.2 部署项目 ······················ 335
  15.5.3 运行项目 ······················ 335

第1章 Java 面向对象开发环境——搭建 Java 开发环境

1.1 第一个 Java 应用程序——Hello World

以开发第一个 Java 应用程序为引导,开启 Java 学习之旅。Java 程序的编写、编译及执行都需要软件环境的支持,所以在开始编写 Java 程序之前,首先要搭建 Java 的开发环境,创建"Hello World"应用程序;然后运行程序并查看运行结果;最后探讨"Hello World"应用程序的执行过程。

1.2 关键技术

1.2.1 Java 语言简介

Java 语言是原 Sun Microsystems 公司(现已被 Oracle 公司收购)推出的 Java 面向对象程序设计语言和 Java 平台的总称,由 James Gosling 及其同事共同研发,并在 1995 年正式推出。Java 最初是为消费类电子产品的嵌入式芯片而设计的,称为 Oak,1995 年正式更名为 Java。

随着互联网技术和 Web 应用技术的飞速发展,Java 被广泛接受。来自 Oracle 公司的官方数据表明,目前全球已有上亿系统是使用 Java 开发的。Java 是一门面向对象的编程语言,它不仅吸收了 C++ 语言的各种优点,还摒弃了 C++ 里难以理解的多继承、指针等概念,因此 Java 语言具有功能强大和简单易用两个特征。

1.2.2 Java 平台

Java 既是一门编程语言,也是一种软件平台。软件平台可以理解为 Java 程序运行的软件环境。Java 软件平台是运行于操作系统之上的,由 Java 虚拟机(Java Virtual Machine,JVM)和 Java 应用编程接口(Java Application Programming Interface,Java API)两部分构成。

Java 虚拟机,负责解释执行 Java 程序。JVM 是操作系统相关的,也就是基于不同的操作系统编写相应的 JVM,JVM 屏蔽了底层平台(如操作系统)的差异,为 Java 程序的平台无关性提供重要支撑。

Java 应用编程接口,是一套独立于操作系统的标准组件的集合,可分为基本部分与扩展部分,按照功能通过包(package)的方式进行组织。

Java 分成 3 种平台，分别如下。

Java SE：Java Platform Standard Edition，即 Java 标准版，Java 语言的核心与基础，它允许开发和部署在桌面、服务器、嵌入式和实时环境中使用的 Java 应用程序。Java SE 既是基础包，同时也包含支持 Java Web 服务开发的类，并为 Java EE 提供基础。

Java EE：Java Platform Enterprise Edition，即 Java 企业版。Java 企业版可以帮助用户开发和部署可移植、健壮、可伸缩且安全的服务器端 Java 应用程序。Java EE 是在 Java SE 的基础上构建的，它提供 Web 服务、组件模型、管理和通信 API，可以用来实现企业级的面向服务体系结构(Service-Oriented Architecture，SOA)和 Web 2.0 应用程序。

Java ME：Java Platform Micro Edition，即 Java 微型版。Java 微型版为在移动设备和嵌入式设备上运行的应用程序提供健壮且灵活的环境。Java ME 包括灵活的用户界面、健壮的安全模型、许多内置的网络协议以及对可以动态下载的联网和离线应用程序的丰富支持。

1.2.3 Java 语言的特点

Sun 公司对 Java 编程语言的解释是：Java 编程语言是一个简单、面向对象、分布式、解释型、健壮、安全与系统无关、可移植、高性能、多线程和动态的语言。

1. Java 语言是简单的

Java 语言的语法与 C 语言和 C++ 语言很接近，使得大多数程序员很容易学习和使用。另一方面，Java 丢弃了 C++ 中很难理解的一些特性，如操作符重载、多重继承等。特别地，Java 语言不使用指针，并提供了自动的垃圾回收机制，使得程序员不必为内存管理而担忧。

2. Java 语言是面向对象的

面向对象是 Java 语言最基本的特性。Java 语言的设计完全是面向对象的，Java 语言提供类、接口和继承等面向对象的特性。为了简单起见，只支持类之间的单继承，但支持接口之间的多继承，并支持类与接口之间的实现机制。

3. Java 语言是分布式的

Java 语言支持网络应用程序的开发，在基本的 Java API 中有一个网络应用编程接口，它提供了用于网络应用编程的类库，如 URL、URLConnection、Socket、ServerSocket 等。因此，Java 应用程序可凭借 URL 打开并访问网络上的对象，其访问方式与访问本地文件系统几乎完全相同。

4. Java 语言是解释型的

Java 源程序(java 文件)在 Java 平台上被编译为字节码格式(class 文件)，而不是通常的机器码格式，然后就可以在实现 Java 平台的任何操作系统环境中运行。在运行时，Java 平台中的 Java 解释器可以对这些字节码进行解释执行，执行过程中需要的类在连接阶段会被载入运行环境。

5. Java 语言是健壮的

Java 的强类型机制、异常处理、垃圾的自动回收机制等是 Java 程序健壮性的重要保证。Java 语言不支持指针，杜绝了内存的非法访问。另外，Java 的安全检查机制也使得 Java 更具健壮性。

6. Java 语言是安全的

Java 语言作为一个面向网络应用的编程语言，其安全性方面自然存在较高的要求。如果没有安全性的保障，就很容易受到恶意代码的攻击。一方面，Java 编译时数据类型检查和自动内存管理可使代码更健壮，以减少内存损坏和漏洞；另一方面，字节码验证可确保代码符合 JVM 规范并防止恶意代码破坏运行环境。同时，通过类加载器可防止不受信任的代码干扰其他 Java 程序的运行。

7. Java 语言是系统无关的

Java 语言是系统无关的，具有跨平台性，这是指 Java 应用程序可以不受计算机硬件和操作系统的约束而在任意计算机环境下正常运行。Java 平台自带的虚拟机很好地实现了跨平台性。Java 源程序代码经过编译后生成的字节码是与平台无关的，是可被 Java 虚拟机识别的一种机器码指令，Java 虚拟机提供了一个从字节码到底层硬件平台及操作系统的屏障，掩盖了底层平台之间的差别，这使得 Java 应用程序可以运行在任何一台具有 Java 虚拟机的机器上。

8. Java 语言是可移植的

Java 语言的可移植性来源于 Java 语言的系统无关性。另外，Java 还严格定义独立于平台的基本数据类型及其运算，Java 数据可以在任何硬件平台上保持一致。

9. Java 语言是高性能的

作为一种解释型编程语言，与那些解释型的高级脚本语言相比，Java 是高性能的。另外，Java 的运行速度随着 JIT(Just-In-Time) 编译器技术的发展也越来越接近于 C++。

10. Java 语言是多线程的

多线程功能使得在一个程序中可以同时执行多个小任务。C 和 C++ 采用单线程体系结构，而 Java 语言支持多线程技术，支持多个线程同时执行，并提供多线程之间的同步机制。多线程技术为应用程序带来了更好的交互性能和实时控制性能。

11. Java 语言是动态的

Java 语言具有一定的动态性。如利用反射机制可以在运行时加载和使用编译期间完全未知的类。Java 语言的动态性使它能够胜任分布式系统环境下的应用，位于各地的类可以自由升级，而不影响原 Java 应用程序的运行。

1.2.4　Java 小程序和应用程序

Java 程序有两种形式：小程序（Applet）和应用程序（Application）。小程序是嵌入 HTML 文件中的 Java 程序，不能独立运行，需要来自 Web 浏览器的大量信息，在浏览器环境中运行并显示，它需要知道何时启动、何时放入浏览器窗口、何处、何时激活、关闭等。Java 小程序可以直接利用浏览器或 AppletViewer 提供的图形用户界面，无须另外构建图形用户界面。

Java 应用程序是能够独立运行的 Java 程序，可以直接在命令行状态下启动运行。相对于小程序直接利用浏览器提供图像用户界面，Java 应用程序必须另外书写专用代码以构建自己的图形用户界面。

1.3　搭建 Java 开发环境的基本步骤

开发 Java 应用程序需要多种工具和一定的软件环境，如 JDK、Eclipse 等。本节主要介绍 Java 开发环境的具体搭建步骤。

1.3.1　JDK 安装与配置

1. 下载与安装 JDK

Java 开发工具包（Java Development Kit，JDK）提供编译、运行 Java 应用程序必需的开发工具。除此，JDK 中还包括 Java 的核心类库。读者可以登录 Oracle 公司的 Java 技术支持网站（http://www.oracle.com/technetwork/java/javase/downloads/index.html）下载 JDK。要根据自己的平台选择合适的 JDK 版本。本书采用 Windows 系统（64 位）下最新版本的 Java SE-- JDK14，本书所有实验都基于该版本。可以分别选择压缩文件包或者安装文件下载。下载成功后分别得到如下文件：压缩文件 jdk-14_windows-x64_bin.zip 或者安装文件 jdk-14_windows-x64_bin.exe 。如果选择压缩文件，则直接解压缩即可；对于安装文件，需要双击安装文件执行安装操作。

JDK 的安装方法很简单，双击可执行文件，单击"下一步"按钮，显示如图 1-1 所示的 JDK 安装界面。选择安装目录，单击"下一步"按钮进行安装，直至安装结束。

安装 JDK 后可得到如下文件路径。

bin：该路径下存放了 JDK 的各种工具命令，如常用的编译工具 Javac、运行工具 Java 等都存放在该路径下。

conf：该路径下存放了 JDK 的相关配置文件。

include：存放一些平台特定的头文件。

jmods：该目录下存放了 JDK 的各种模块。

图 1-1 Java 安装界面

legal：该目录下包含了 JDK 各模块的授权文档。

lib：该路径下存放的是 JDK 工具的一些补充 JAR 包。

2. 配置环境变量

安装结束后，需要配置操作系统环境变量。在桌面上选择"此电脑"图标并右击，选择"属性"选项，显示"控制面板|系统和安全|系统"界面，选择右侧"高级系统设置"选项，弹出如图 1-2 所示的对话框，单击"环境变量"按钮，显示如图 1-3 所示的设置环境变量的对话框。可以在用户变量或者系统变量中新建和编辑如图 1-4 和图 1-5 所示的环境变量，其中 C:\Program Files\Java\jdk-14 是 JDK 的安装目录。

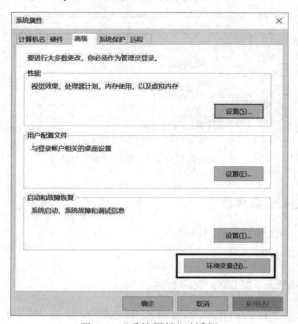

图 1-2 "系统属性"对话框

图 1-3 "环境变量"对话框

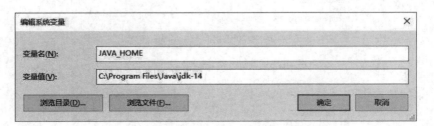

图 1-4 编辑环境变量 JAVA_HOME

3. 测试 JDK 是否安装成功

执行"开始"|"运行"命令,输入 cmd,再输入命令 java -version,出现以下信息则说明已经成功安装及配置 JDK。

4. 利用文本编辑器开发 Java 应用程序

(1) 创建源程序文件:利用文本编辑器生成名为 HelloWorld.java 的源程序文件。源代码如下所示。

第1章 Java 面向对象开发环境——搭建 Java 开发环境

图 1-5　编辑环境变量 Path

```
01 /*
02  * 第 1 章案例
03  */
04 /**
05  * HelloWorld:控制台输出 Hello World。
06  * @author yulp
07  *
08  */
09 public class HelloWorld {                              //类的定义
10     public static void main(String[] args) {           //主方法的定义
11         System.out.println("Hello World.");            //通过标准输出设备输出字符串
12     }
13 }
```

(2) 代码说明。

① 注释。

为了提高程序的可读性和可维护性,通常需要在源程序中包含必要的注释信息。Java 语言中有以下 3 种注释语句。

01~03 代码行:位于"/＊"和"＊/"之间的字符称为多行注释。

09~11 代码行:"//"符号开始的字符称为单行注释。

04~08 代码行:位于"/＊＊"与"＊/"之间的字符称为文档注释,这部分注释内容可以通过 javadoc 命令导出生成 java 文档。

② 类的声明。

格式如下:

public class 类名{
...
}

定义公有类。Java 语言规定一个 Java 源文件中可以包含若干个类定义,但是每个文件仅能包含一个公有类,并且与源文件同名。

③ 主方法。

HelloWorld 类中定义了主方法 main(),main()方法是 Java 应用程序执行的入口方法。main()方法的定义格式如下:

public static void main(String[] args)

修饰符 public 指明 main()方法具有公有的访问权限;修饰符 static 指明 main()方法是一个静态方法。静态方法可以通过类直接调用,而无须实例化对象;关键字 void 表示 main()方法没有返回值;字符串数组 args 是 main()方法的形式参数。通过 args 形参,Java 应用程序可以在运行时接收系统提供的字符串数据。

本例主方法通过标准输出对象 out 调用输出方法 println(),输出字符串。

(3) 编译:在命令行窗口下输入如下命令行。

Javac HelloWorld.java

当前目录下生成名为 HelloWorld.class 的字节码文件。

(4) 运行:在命令行窗口下输入如下命令行。

Java HelloWorld

得到输出:Hello World。

1.3.2 深入理解"HelloWorld"应用程序的执行

如前所述,Java 程序的执行过程分为如图 1-6 所示的两个步骤。

第一步:将 java 源码(java 文件)通过编译器(javac.exe)编译成字节码文件(class 文件)。

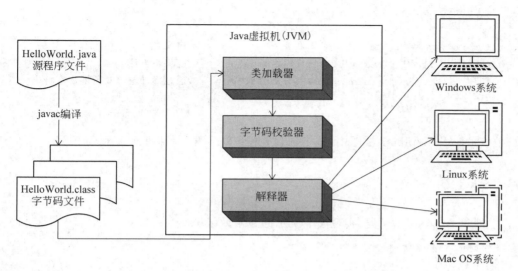

图 1-6　Java 程序执行过程示意

第二步：将字节码文件通过 java.exe 在 JVM 上执行，输出运行结果。

源程序文件(java)经过编译之后生成的字节码文件(class)是一种跨平台的伪代码。在 JVM 中执行应用程序时，首先由类加载器(ClassLoader)和它的子类实现类的加载，也就是指把类的 class 文件读入内存，然后产生与所加载类对应的 Class 对象。加载完成后，还需要经过字节码校验器进行代码的合法性与安全性检测，只有通过检测的代码才可以在解释器上运行。

Java 虚拟机 JVM 是平台相关的。平台无关的字节码文件在解释器上解释执行时，需要经过不同平台的 JVM 进行解析，JVM 会将字节码转换成具体平台上的机器指令。也正是因为存在 JVM，才能够实现 Java 应用程序"一次编译，到处运行"功能。

1.3.3　Eclipse 软件包的下载与安装

Eclipse 是一款开源的、基于 Java 的可扩展的集成开发平台。可以直接登录 Eclipse 的官方网站(http://www.eclipse.org/downloads/eclipse-packages/)下载 Eclipse 集成开发环境 Eclipse IDE for Java Developers。要根据自己的平台选择合适的 Eclipse 版本。下载成功后，会得到压缩文件 eclipse-java-2020-03-R-win32-x86_64.zip。Eclipse 无须安装，解压缩后双击 eclipse.exe 文件即可运行。

启动 Eclipse 后，首先需要设置项目的工作区。进入 Eclipse 之后，选择 File|New|Java Project，创建名为 chap01 的 Java 项目。在 Package Explorer 视图中，选中项目，选择 File|New|class，在项目中新建名为 HelloWorld 的类。

```
package chap01;
/**
 * HelloWorld:控制台输出 Hello World。
```

```
 * @author yulp
 */
public class HelloWorld {                        //定义类
    public static void main(String[] args) {     //定义主方法
        System.out.println("Hello World.");      //通过标准输出对象输出字符串
    }
}
```

在 Package Explorer 视图中,选中项目,选择 Run|Run,即可运行 Java 项目。

1.4 练一练

1. 创建 Accountting 类,实现如下所示的显示效果。

```
*****************************
*                           *
*      欢迎使用家庭记账系统   *
*                           *
*****************************
* 1. 收支明细               *
* 2. 登记收入               *
* 3. 登记支出               *
* 4. 退出系统               *
*****************************
```

2. 创建 AccountTest.java 文件,详细代码如下。

```
public class AccountTest {
    public static void main(String[] args) {
        Account account = new Account();
        account.printAccount();
    }
}
class Account{
    void printAccount() {
        System.out.println("print Account.");
    }
}
```

编译并运行 AccountTest。在项目文件夹中查找生成的字节码文件,并分析程序的功能。

第 2 章 基本数据类型及运算——身体质量指数计算器

2.1 身体质量指数计算器

身体质量指数(Body Mass Index,BMI)是目前国际上通用的衡量人体胖瘦程度以及是否健康的标准。身体质量指数的计算公式为

$$BMI=\frac{体重(kg)}{身高(m)^2}$$

我国参考世界卫生组织(WHO)的标准制定了中国 BMI 参考标准,如表 2-1 所示。

表 2-1 中国 BMI 参考标准

BMI 范围	分 类
BMI<18.5	偏瘦
18.5≤BMI<24.0	正常体重
24.0≤BMI<27.0	偏胖
27.0≤BMI<30.0	肥胖
BMI≥30.0	重度肥胖

1. 输出 BMI 计算器主菜单

```
**********************
   身体质量指数（BMI）
1.计算BMI
2.查看BMI范围
**********************
请输入您的选择（1-2）
```

2. 计算 BMI

```
请输入您的身高（cm）
160

请输入您的体重（kg）
60
您的身体质量指数BMI为：23.44
```

3. 查看 BMI 范围

BMI 分类	BMI 范围
偏瘦	<18.5
正常	18.5~23.9
偏胖	24~26.9
肥胖	27~29.9
重度肥胖	>=30

2.2 程序设计思路

程序设计需要按照一定的方法进行，这样在开发程序时才能事半功倍。IPO 模式是程序设计的一种基本模式：

I：Input，程序的输入。
P：Process，程序的主要运算及处理。
O：Output，程序的输出。

下面按照 IPO 模式分析身体质量指数计算器的程序设计思路。

1. 输入

输入是一个程序的开始，通常包括：用户通过程序界面或者控制台手工输入、文件输入、网络输入、随机数据输入等形式。BMI 计算器可以采用通过控制台手工输入数据的方式。

2. 处理

计算机处理的对象是数据，而数据在计算机中是以某种特定的形式存在的，所以处理数据首先要解决数据的表示及存储问题，然后按照某种方式表示的身高与体重数据就可以根据公式进行运算，最终得到 BMI 指数。

3. 输出

IPO 的最后一个环节是通过程序的输出展示运算结果，通常包括屏幕显示输出、文件输出、网络输出等形式。BMI 计算器可以采用屏幕显示输出的方式展示运算结果。

2.3 关键技术

2.3.1 Java 标识符

用来标识变量、常量、类、方法等实体名字的字符序列称为标识符。简单地说，标识符就是一个名字。Java 标识符的命名规则如下。

- 所有的标识符都应该以字母、美元符($)或者下画线(_)开始。

- 首字符之后可以是字母、美元符（$）、下画线（_）或数字的任何字符组合。
- 关键字不能用作标识符（关键字见 2.3.2 节）。
- 标识符是大小写敏感的。
- 标识符的长度不受限制。
- "字母"的范围不仅包括 A~Z 或者 a~z 等英文字母，还包括 Unicode 字符集中的字符，如汉字等。

为了提高代码的可读性，在 Java 语言中提倡规范化的标识符命名，常用的命名规范约定：包名使用小写字母；类名和接口名通常由具有含义的单词构成，所有单词的首字母大写，如类 Student；成员变量与方法名通常由具有含义的单词组成，第一个单词首字母小写，其他单词的首字母大写，如 displayStudent()方法；变量名通常全部使用小写；常量名的全部字母均为大写，最好使用下画线分割单词，如 STUDENT_COUNT。

2.3.2 关键字

关键字（Keyword）是 Java 语言保留的具有特定含义的英文单词。每个关键字都具有特定的含义，不能再作为标识符使用。Java 定义了如下 50 个关键字。

abstract	continue	for	new	switch	assert	default
goto	package	synchronized	boolean	do	if	private
this	break	double	implements	protected	throw	byte
else	import	public	throws	case	enum	instanceof
return	transient	catch	extends	int	short	try
char	final	interface	static	void	class	finally
long	strictfp	volatile	const	float	native	super
while						

注意：虽然 null、true 及 false 不是关键字，但是它们也不能作为标识符使用。

2.3.3 Java 的数据类型

数据类型是具有相同逻辑意义的一组值的集合以及定义在这个集合上的一组操作。Java 语言的数据类型包括基本数据类型与引用数据类型，其中，引用数据类型非常类似于 C/C++ 的指针变量类型，例如对象、数组都是引用数据类型。基本数据类型的定义如图 2-1 所示。

1. 布尔类型

布尔类型（boolean）又称逻辑类型。
- 常量：true、false。
- 变量：使用关键字 boolean 声明布尔类型的变量。

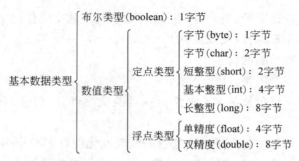

图 2-1 Java 的基本数据类型

2. 整数类型

Java 语言中,基本的整型数据类型有 byte、short、int、long 四种,用于定义不同存储空间的数据。Java 整数类型的定义如表 2-2 所示。

表 2-2 整数类型的定义

整数类型	长度(字节数)	取值范围
byte	1	−128〜127
short	2	−32768〜32767
int	4	−2147483648〜2147483647
long	8	−9223372036854775808〜9223372036854775807

1) 常量

整数型常量有以下 3 种表示形式:

(1) 十进制整数,如 12、−34 等。

(2) 八进制整数,以 0 开头的整数,如 056。

(3) 十六进制整数,以 0x 开头的整数,如 0x25、0xa2。

另外,后缀 L 或者 l 表示 long 型常量,如 322L。

2) 变量

使用 byte、short、int、long 关键字声明整数型变量,声明时可以同时赋初值,如 int counter=100;要注意不同整数类型的取值范围。

3. 浮点类型

浮点类型是包含小数部分的数值。Java 语言中的浮点类型数据根据取值范围的不同分为 float(单精度)和 double(双精度),如表 2-3 所示。

表 2-3 Java 浮点类型的定义

浮点类型	长度(字节数)	取值范围
float	4	−3.40E+38〜+3.40E+38
double	8	−1.79E+308〜+1.79E+308

1) 常量

浮点类型常量有以下 2 种表示形式。

（1）标准记数法，由整数部分、小数点及小数部分组成，如 0.12。

（2）科学记数法，由十进制整数、小数点、小数和指数部分构成，指数部分由 E 或者 e 开头，后面是带正负号的整数，如 2.33E+23。

另外，后缀 F 或者 f 表示 float 型常量，如 3.21F；后缀 D 或者 d 表示 double 型常量，如 3.21D。

2) 变量

使用 float、double 关键字声明浮点类型变量，声明时可以同时赋初值，如 float sum=0.5F、double average=0.5。

注意：浮点类型的常量默认是 double 类型的。

4. 字符类型

字符类型(char)数据是由单引号(' ')定界的单个字符，该字符是 16 位的 Unicode 字符，支持 0~65535 的字符编码，可以表示英文、汉字、日文与韩文等符号。

1) 常量

字符类型常量有以下 2 种表示形式：

（1）单引号定界的单个字符，如'A'、'中'。

（2）单引号定界的 Unicode 编码，其中 Unicode 编码以"\u"开头，后面是十六进制表示的 Unicode 编码。如'\u0061'表示 Unicode 编码为 0x61 的字符，也就是字符'a'。

另外，以'\'开头，后面跟一个字母表示的特定字符称为转义字符。Java 转义字符的定义如表 2-4 所示。

表 2-4　Java 转义字符的定义

转义字符	含义	转义字符	含义
\'	单引号	\r	回车符
\"	双引号	\\	反斜杠
\b	退格	\n	换行符
\t	制表符		

2) 变量

使用 char 关键字声明字符型变量，如：char c='a'。

2.3.4　基本数据类型的转换

在表达式中，不同类型的数据需要进行类型转换。Java 程序中的类型转换分为隐式类型转换和显式类型转换两种形式。

隐式类型转换：在进行数据运算时，如果参与运算的操作数类型不一致，则系统会自动将精度较低的类型转换为精度较高(即占用内存较多,取值范围较大)的类型,称为隐式数据类型转换。Java 的基本数据类型按照精度从低到高依次排列如下:

```
byte short char int long float double
```

显式(强制)类型转换:需要在表达式的前面指明所需要转换的类型,系统将按这一要求把某种类型强制转换为指定的类型。

```
(<类型名>)<表达式>
```

2.3.5 运算符与表达式

Java 语言提供了多种运算符,按照参与运算的操作数的个数,可以将运算符分为单目运算、双目运算及三目运算;按照运算符的功能可分为算术运算符、赋值运算符、关系运算符、逻辑运算符、位运算符及条件运算符。运算符将操作数按照 Java 语法规范连接在一起即组成表达式。

1. 算术运算符

基本的算术运算符包括: ＋、－、*、/、％、＋＋、－－。

运算符/:若操作数均为整数,则结果也为整数;若有一个操作数为浮点类型,则结果就是浮点类型。

运算符＋:Java 中允许对字符串数据进行"＋"运算,以实现字符串的连接。如"123"＋"45"将得到字符串"12345"。

运算符％:其操作数既可以是整型数据,也可以是浮点类型数据。"％"运算结果的符号与第一个操作数的符号相同。

＋＋(自增运算符)、－－(自减运算符):根据运算符的位置不同可以分为前置自增运算(＋＋i)、前置自减运算(－－i)、后置自增运算(i＋＋)、后置自减运算(i－－)。其中,前置自增、自减运算先进行自增或者自减运算,再进行表达式运算;而后置自增、自减运算先进行表达式运算,再进行自增或者自减运算。

2. 关系运算符

关系运算符包括:<、<=、>、>=、==、!=。用于两个操作数的比较,运算结果是 boolean 型。

==、!=运算符:用于比较基本数据类型数据时,比较的是操作数的值是否相等;对于引用类型时,比较的是操作数的引用是否相同,即比较两个操作数是否是同一对象的引用,而不对引用的内容进行比较。

对于两个浮点类型的数据,通常不建议进行"=="操作。因为浮点数据在计算机内存表示时都是存在精度误差的。如:

```
0.1+1.0==1.1          //运算结果为 false
```

两个浮点数的比较通常通过两个浮点数差的绝对值与预先设定的精度判定。如：

```
Math.abs(fdata1-fdata2)<1e-6        //两个浮点类型数据 fdata1,fdata2
```

其中，1e-6 就是预设的比较精度数据，Math.abs()是 Java 中定义的绝对值方法。

3. 逻辑运算符

逻辑运算符包括：&、|、&&、||、!。逻辑运算的操作数和运算结果都是 boolean 型数据。表 2-5 给出了逻辑运算的真值表。其中，oper1 与 oper2 表示 boolean 类型的操作数。

表 2-5 逻辑运算的真值表

oper1	oper2	oper1&&oper2 oper1&oper2	oper1||oper2 oper1|oper2	!oper1
true	true	true	true	false
true	false	false	true	false
false	true	false	true	true
false	false	false	false	true

由表 2-5 可知，逻辑运算符"&&"与"&"的逻辑运算结果相同。同样，逻辑运算符"||"与"|"也具有相同的逻辑结果，但是其运算过程是不同的。其中，"&&"和"||"也称短路逻辑运算符，这是因为当第一个操作数 oper1 为 false 时，"&&"运算就不会再计算第二个操作数 oper2，直接给出表达式 oper1&&oper2 的运算结果为 false；同理，当 oper1 为 true 时，"||"运算也将短路第二个操作数 oper2，直接给出表达式 oper1||oper2 的运算结果为 true。而逻辑运算"&"和"|"则不遵循短路原则，即无论第一个操作数的取值为 true 还是 false，都将计算第二个操作数。

例如，已知变量 x=1：

```
(2>3) &&(x=2)>1    //运算之后,表达式的结果为 false,x= 1
(2>3) &(x=2)>1     //运算之后,表达式的结果为 false,因为执行了第二个操作数,所以 x=2
```

同理：

```
(2<3)||(x=2)>1     //运算之后,表达式的结果为 true,x=1
(2<3)|(x=2)>1      //运算之后,表达式的结果为 true,x= 2
```

4. 位运算符

Java 的位运算符包括：&、|、~、^、>>、<<、>>>，应用于 int、long、short、char 和 byte 等类型。

位运算符作用在操作数的所有位上，并且按位运算。假设 op1=-1,op2=17，它们的二进制格式表示如下：

```
op1=11111111 11111111 11111111 11111111
op2=10000000 00000000 00000000 00010001
```

在计算机中,数据采用补码方式表示,最高位对应符号位,0 代表正数,1 代表负数。位运算 op1 & op2＝00000000 00000000 00000000 00010001。

5. 赋值运算符

赋值运算符包括"＝"及复合赋值运算符。表 2-6 给出了赋值运算符的含义和用法。

表 2-6　赋值运算符

运算符	含　　义	用　　法
＝	基本赋值运算符,计算右操作数的值,并赋值给左侧变量	z＝x＋y 计算 x＋y,将结果赋值给 z
＋＝	加赋值运算符,计算左、右操作数之和,并赋值给左侧变量	x＋＝y 计算 x＋y,将结果赋值给,等价于 x＝x＋y
－＝	减赋值运算符,计算左、右操作数之差,并赋值给左侧变量	x－＝y 计算 x－y,将结果赋值给 x,等价于 x＝x－y
＊＝	乘赋值运算符,计算左、右操作数之积,并赋值给左侧变量	x＊＝y 计算 x＊y,将结果赋值给 x,等价于 x＝x＊y
/＝	除赋值运算符,将左、右操作数相除,把结果赋值给左侧变量	x/＝y 计算 x/y,将结果赋值给 x,等价于 x＝x/y
%＝	求余赋值运算符,将左、右操作数求余,把结果赋值给左侧变量	x%＝y 计算 x%y,将结果赋值给 x,等价于 x＝x%y
<<＝	左移位赋值运算符,将左操作数左移右操作数指定的位数,并将移位之后的结果赋值给左侧变量	x<<＝2 等价,x＝x<<2
>>＝	右移位赋值运算符,将左操作数右移右操作数指定的位数,并将移位之后的结果赋值给左侧变量	x>>＝2 等价于 x＝x>>2
&＝	按位与赋值运算符,计算左、右操作数按位与运算,并将结果赋值给左侧变量	x&＝y 等价于 x＝x&y
|＝	按位或赋值运算符,计算左、右操作数按位或运算,并将结果赋值给左侧变量	x|＝y 等价于 x＝x|y
^＝	按位异或赋值运算符,计算左、右操作数按异或运算,并将结果赋值给左侧变量	x^＝y 等价于 x＝x^y

6. 条件运算符(?:)

Java 的条件运算符是三目运算,形如:

x?y:z

该运算符首先计算 x 的值,若 x 的值为 true,则计算 y,并且将 y 的值作为条件表达式的值;否则计算 z,并将 z 的值作为条件表达式的值。

7. instanceof 运算符

双目运算实例运算符,用于判断指定对象是否是一个特定类型(类类型或接口类型)。例如:

```
obj instanceof String
```

其中,obj 是一对象实例,若 obj 是 String 类型实例,则运算结果为 true,否则为 false。

8. Java 表达式

将 Java 运算符与操作数按照 Java 语法规则连接起来,就构成了 Java 表达式。其中,操作数可以是变量、常量或者方法调用。当表达式中出现多个运算符时,如何判断运算符的执行顺序呢?这就涉及运算符的优先级问题,应先计算优先级高的运算。具有多个相同优先级的运算符进行运算时,就需要考虑运算符的结合性,包括左结合(从左到右运算)、右结合(从右到左运算)。

除此,表达式中可以采用圆括号改变运算符的运算次序,将优先计算圆括号中的表达式。编写程序时,遇到比较复杂的表达式应尽量使用圆括号实现预期的运算顺序,以增加程序的可读性。

Java 运算符的优先级和结合性如表 2-7 所示。

表 2-7 Java 运算符的优先级与结合性

优先级	运算符	结合性	描述
1(高)	[ ] ( )	左结合	成员、括号
2	++、--	左结合	后置自增、后置自减
	++、--、+、-、~、!	右结合	前置自增、前置自减、取正、取负、求反
3	*、/、%	左结合	乘、除、求余运算符
4	+、-	左结合	加、减
5	>>、>>>、>>	左结合	移位
6	>、>=、<、<= instanceof	左结合	关系、实例
7	==、!=	左结合	关系
8	&	左结合	按位与
9	^	左结合	按位异或
10	|	左结合	按位或
11	&&	左结合	逻辑与
12	||	左结合	逻辑或
13	?:	左结合	条件
14	=、+=、-=、*=、/=、%=、>>=、<<=、&=、^=、|=	右结合	赋值及复合赋值
15(低)	,	左结合	逗号运算

2.3.6　简单数据的输入与输出

1. 利用文本扫描器类 Scanner 实现数据的输入

Scanner 类的常用方法为

```
public Scanner(InputStream source)
```

Scanner 构造方法用于构建一个扫描器对象。如：

```
Scanner sc=new Scanner(System.in);
```

对标准输入对象 in 进行封装，构建扫描器对象 sc。

```
public boolean hasNextXxx()
```

判断是否存在下一个指定类型的输入数据，其中"Xxx"代表基本的数据类型，如 long、Int 等。

```
public Xxx nextXxx()
```

获取下一个指定类型的输入项。通常数据之间采用空格、Tab 键及回车键进行间隔。

2. 利用 System 类实现数据的输出

```
stem.out.println(字符串)              //输出字符串并换行
System.out.print(字符串)              //输出字符串
System.out.printf(格式控制字符串,表达式1,表达式2,…,表达式n)    //格式化输出
```

其中，格式控制字符串由普通字符和格式控制符两部分组成。进行数据输出时，普通字符原样输出，格式控制符用于控制相关表达式的输出格式。常用的格式控制符如表 2-8 所示。

表 2-8　格式控制符的定义

格式控制符	描述
%d	以十进制形式输出整型数据
%c	输出 char 型数据
%s	输出字符串数据
%f	以十进制形式输出浮点数，最多保留 6 位小数
%e	以科学记数法形式输出浮点数
%0md	控制输出的整数占 m 位，不足 m 位用 0 填充
%m.nf	控制输出的浮点数占 m 位，其中小数位占 n

2.4　身体质量指数计算器设计步骤

1. 程序输入

身体质量指数的计算需要两个输入数据：身高与体重。根据 BMI 计算公式，体重、身高的单位分别为"千克"与"米"。程序设计时，为保证 BMI 的计算精度，将身高的单位定义为"厘米"。两个输入数据定义为 double 类型，利用 Scanner 类提供的 nextDouble() 方法获取用户通过控制台手工输入的数据。

2. 设计算法

身体质量指数的计算公式为

$$\mathrm{BMI} = \frac{体重(\mathrm{kg})}{身高(\mathrm{cm})^2 \times 10^{-4}}$$

3. 程序输出

利用 System 类实现程序菜单及 BMI 运算结果的输出。
身体质量指数计算器的源代码如下。

```java
package chap02;
import java.util.Scanner;
public class BMI {
public static void main(String[] args) {
    System.out.printf("*********************\n");
    System.out.printf("%15s\n", "身体质量指数(BMI)");
    System.out.println("1.计算 BMI");
    System.out.println("2.查看 BMI 范围");
    System.out.printf("*********************\n");
    System.out.print("请输入您的选择(1-2)");
    Scanner sc = new Scanner(System.in);
    int choice = sc.nextInt();
    if (choice == 1) {
        System.out.println("\n 请输入您的身高(cm)");
        double height = sc.nextDouble();
        System.out.println("\n 请输入您的体重(kg)");
        double weight = sc.nextDouble();
        double bmi = weight/(height * height * 1e-4);
        System.out.printf("您的身体质量指数 BMI 为：%.2f\n", bmi);
    } else if (choice == 2) {
        System.out.printf("%s\t\t%s\n", "BMI 分类", "BMI 范围");
        System.out.printf("%s\t\t%s\n", "偏瘦", "<18.5");
        System.out.printf("%s\t\t%s\n", "正常", "18.5~23.9");
```

```
            System.out.printf("%s\t\t%s\n", "偏胖", "24~26.9");
            System.out.printf("%s\t\t%s\n", "肥胖", "27~29.9");
            System.out.printf("%s\t\t%s\n", "重度肥胖", ">= 30");
        } else
            System.out.println("输入错误。");
        sc.close();
    }
}
```

2.5 练一练

1. 输出 Java 基本数据类型的取值范围。

提示：Java 语言为基本数据类型提供了相应的包装类，分别是 Boolean、Byte、Character、Short、Integer、Long、Float、Double。包装类中封装了静态成员变量 MIN_VALUE 与 MAX_VALUE，分别表达基本数据类型所能表示的最小值与最大值。

2. 输出汉字及其 Unicode 编码，并分析编码的排列规则。

提示：如首先输出汉字"拉"的 Unicode 编码，然后输出 Unicode 编码表位于"拉"之后的几个汉字，分析 Unicode 编码的排列规则。

3. Java 语言的基本数据类型、运算符与 C 语言相比有哪些不同之处？

第 3 章　结构化程序设计——设计图书 ISBN 校验器

3.1　图书 ISBN 校验器

国际标准书号(International Standard Book Number,ISBN)是专门为识别图书等文献而设计的国际编号。存在两种形式的 ISBN 编码：10 位 ISBN 编码与 13 位 ISBN 编码。

10 位 ISBN 编码的组成方式：组号-出版者号-书序号-校验码，例如 7-302-08599-4。组号是国家、地区、语言的代号；出版者号是出版社代号，由其隶属的国家或地区的 ISBN 中心分配；书序号由出版社定义其发行图书的编号；最末位的校验码能够校验出 ISBN 号是否正确。10 位 ISBN 编码的校验码满足如下公式：

$$C_{10} = 11 - \mathrm{MOD}\left(\sum_{i=1}^{9}(11-i) \times C_i, 11\right)$$

$C_1 \sim C_{10}$ 分别表示 1 位 ISBN 编码，其中，$C_1 \sim C_9$ 取值为 0~9 的数字，C_{10} 可以取值字母 X 及数字 0~9，如果校验码 $C_1=10$，则用 X 表示。

13 位 ISBN 编码的组成方式：前缀码-组号-出版者号-书序号-校验码，例如 978-7302-08599-7。在 10 位编码的基础上增加了前缀码：978 或 979。校验码满足如下公式：

$$C_{13} = 10 - \mathrm{MOD}\left(\sum_{\substack{i \leqslant 12 \\ \text{且 i 为奇数}}} C_i + 3 \times \sum_{\substack{i \leqslant 12 \\ \text{且 i 为偶数}}} C_i, 10\right)$$

如果计算得到 $C_{13}=10$，则令 $C_{13}=0$，以保证最末位的校验码取值为 0~9 的数字。

3.2　程序设计思路

依然采用 IPO 程序设计模式。
(1) 程序的输入(I)：以字符串的形式输入待检测的 ISBN 编码。
(2) 程序处理过程(P)：这部分包括程序的核心算法，需要依次完成以下操作：
① 需要获取输入的 ISBN 编码的长度，进而根据长度选择不同的公式计算校验码；
② 获取输入编码的第 i 位数字，即 C_i；
③ 根据公式计算累加和及余数。
(3) 程序的输出(O)：给出判断结果，即是否为合法的 ISBN 校验码。

3.3 关键技术

为了使程序具有良好的结构，使程序易于设计、易于理解，通常高级语言都会提供 3 种基本的流程控制结构：顺序结构、分支结构和循环结构。顺序结构按照语句的书写顺序执行程序；分支结构根据条件的取值选择性地执行代码分支；循环结构根据循环条件重复执行代码段。

3.3.1 顺序结构

顺序结构指计算机按照语句的编写次序执行，没有分支和跳转语句。常见的顺序结构包括：表达式语句、复合语句及空语句。

1. 表达式语句

在表达式的末尾加上分号即可构成表达式语句，例如：count=1。

2. 复合语句

使用大括号把若干语句和声明组合到一起即可构成复合语句。

3. 空语句

仅由分号构成，不执行任何操作。

3.3.2 分支结构

Java 语言中常用的分支结构包括 if 语句及 switch 语句。

1. if 语句

if 语句根据逻辑表达式的取值决定程序的执行分支。if 语句有如下 3 种格式。
格式 1：单分支语句。

```
if(逻辑表达式){
    语句序列;
}
```

流程图如图 3-1 所示。
格式 2：双分支语句。

```
if(逻辑表达式){
    语句序列 1;
} else {
```

    语句序列 2;
}

流程图如图 3-2 所示。

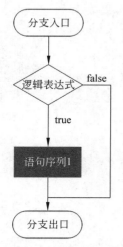

图 3-1　if 语句单分支流程图

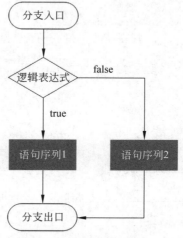

图 3-2　if 语句双分支流程图

格式 3：多分支语句。

```
if (逻辑表达式 1){
    语句序列 1;
} else if(逻辑表达式 2){
    语句序列 2;
} …
else {
    语句序列 n+1
}
```

流程图如图 3-3 所示。

3 种格式是相通的，如果格式 2 省略 else 子句，它就变成了格式 1；同理，格式 3 中省略 else if 子句就变会成格式 2。一般而言，如果语句序列中仅包含一行语句，则可以省略花括号；但是为了增强程序的可读性，通常不建议省略花括号。

分析下面的例 3-1，根据输入的成绩判断所属的成绩等级。

例 3-1

```java
import java.util.Scanner;
public class ScoreGroup {
    public static void main(String[] args) {
        Scanner sc = new Scanner(System.in);
        System.out.println("请输入成绩:");
        intscore = sc.nextInt();
        if(score>=0 &&score<=59) {
```

```java
        System.out.println("不及格");
    }else if(score<=69) {                    ①
        System.out.println("及格");
    }else if(score<=89) {
        System.out.println("良");
    }else if(score<=100) {
        System.out.println("优秀");
    }else {
        System.out.println("输入信息有误");
    }
    sc.close();
    }
}
```

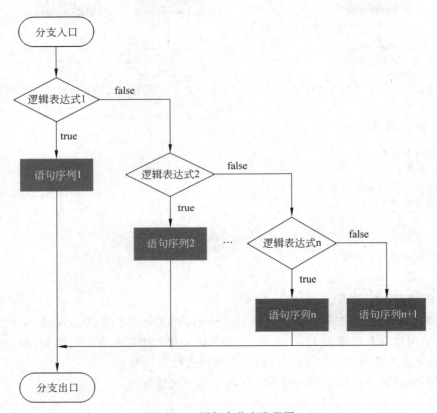

图 3-3　if 语句多分支流程图

表面上看起来，上面的程序没有问题，分别输入成绩 50、65、79、94、120 时，可以得到预期的输出结果。但是当输入成绩为 −4 时，运行上面的程序，会得到及格的成绩等级，显然程序出现了问题。经过分析发现，主要原因在于 else 子句虽然后面仅包含一个条件，但是实际上 else 子句还隐含了一个条件：对前面条件取反。如例 3-1 的代码行①，条件是 score<=69，实际上这里隐含的条件是：!(score>=0 && score<=59) && score

<=69,等价于 score<0 || (score>59 && score <=69)。显然,当输入成绩为-4 时,程序出现了问题。

为了得到正确的程序逻辑,把例 3-1 修改为如下形式。

```java
import java.util.Scanner;
public class ScoreGroup {
    public static void main(String[] args) {
        Scanner sc = new Scanner(System.in);
        System.out.println("请输入成绩:");
        int score = sc.nextInt();
        if(score<0 || score>100) {
            System.out.println("输入信息有误");
        }else if(score>=90) {
            System.out.println("优秀");
        }else if(score>=70) {
            System.out.println("良");
        }else if(score>=60) {
            System.out.println("及格");
        }else {
            System.out.println("不及格");
        }
        sc.close();
    }
}
```

2. switch 语句

语法格式如下:

```
switch (表达式){
    case 值 1:语句序列 1;break;
    case 值 2:语句序列 2;break;
    ...
    case 值 n:语句序列 n;break;
    default: 语句序列 n+1;
}
```

流程表示如图 3-4 所示。

switch 中的表达式只能是 byte、short、char、int 及枚举类型。Java 7 之后,增加了 String 类型的表达式。

switch 语句的执行过程:计算表达式将表达式的值按从上至下的顺序依次与 case 子句的常量进行比较。当表达式的值与 case 子句的常量相等时,执行其后的语句序列,直到遇到 break 或者 switch 语句执行结束。若没有与表达式的值相同的 case 常量,则执行 default 子句对应的语句序列,若程序省略 default 子句,则结束 switch 语句。

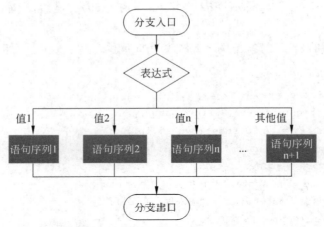

图 3-4　switch 语句多分支流程图

3.3.3　循环结构

Java 语言常用的 3 种循环结构为 while 循环、do-while 循环及 for 循环。

1. while 循环

语法格式如下：

```
初始化;
while(逻辑表达式){
    语句序列;
}
```

流程表示如图 3-5 所示。

while 循环的执行过程如下：

① 执行初始化操作；

② 计算逻辑表达式的值，若值为 true，则转到③，否则转到④；

③ 执行语句序列，转到②；

④ 循环出口，while 语句结束。

使用 while 循环时，一定要保证作为循环条件的逻辑表达式可以变成 false，否则这个循环就会成为死循环，程序无法结束。

2. do-while 循环

语法格式如下：

```
初始化;
do
{
```

    语句序列；
} while(逻辑表达式);

流程表示如图 3-6 所示。

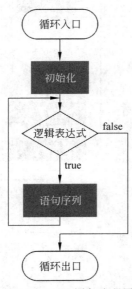

图 3-5 while 语句流程图

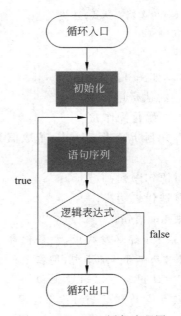

图 3-6 do-while 语句流程图

do-while 循环与 while 循环相比主要的区别在于：while 循环是先判断逻辑表达式后再执行循环语句序列；而 do-while 循环是先执行循环语句序列，然后判断逻辑表达式，如果逻辑表达式取值为 true，则继续执行语句序列，否则中止循环。

初学者使用 do-while 循环时容易出现的错误如下：

① 忽略 while(逻辑表达式)后的分号，造成编译错误；

② 在循环内部，声明循环变量，例如如下代码段。

```
do {
    System.out.println("请输入一个 1~10 之间的整数");
    int guess = sc.nextInt();
    if(guess<truth)
        System.out.println("太小了");
    else if(guess>truth)
        System.out.println("太大了");
}while(guess!=truth);
```

循环内部声明的局部变量 guess 的作用域仅限于大括号定界的语句序列，不包括 while 子句后面的逻辑表达式，所以上述程序会提示"cannot be resolved to a variable"错误。

3. for 循环

语法格式如下：

```
for(初始化;逻辑表达式;迭代表达式){
    语句序列;
}
```

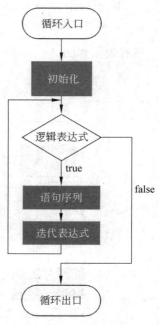

图 3-7　for 语句流程图

流程表示如图 3-7 所示。

for 语句的执行过程如下：
① 执行初始化操作；
② 计算并判断逻辑表达式，若取值为 true，则转到③，否则转到⑤；
③ 执行语句序列；
④ 计算迭代表达式；
⑤ 结束 for 语句。

注意：在 for 循环结构中，初始化部分定义的局部变量的作用范围仅局限于 for 语句，即在 for 语句体之外，该局部变量将不可以使用。可以通过把初始化部分的代码放置在 for 语句之前扩大局部变量的作用范围。

3.3.4　循环控制结构

常用的循环控制结构包括 break 语句、continue 语句及 return 语句。

1. break 语句

break 语句用于结束循环。以 for 循环结构为例，语法定义如下：

```
for(初始化;逻辑表达式 1;迭代表达式){
    语句序列 1;
    if(逻辑表达式 2)break;
    语句序列 2;
}
```

流程控制如图 3-8 所示。

注意：对于循环嵌套结构，break 仅结束其所在的循环结构。

2. continue 语句

continue 语句的作用是忽略本次循环剩余的语句，继续执行下一次循环。以 for 循环结构为例，语法定义如下：

```
for(初始化;逻辑表达式 1;迭代表达式){
```

```
    语句序列 1;
    if(逻辑表达式 2)continue;
        语句序列 2;
}
```

流程控制如图 3-9 所示。

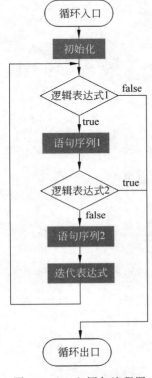

图 3-8　break 语句流程图

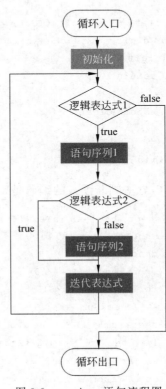

图 3-9　continue 语句流程图

注意：当逻辑表达式 2 取值为 true 时，流程控制将短路语句序列 2，会执行迭代表达式，继续下一次循环。

3. return 语句

结束当前方法的调用，返回主调方法。

3.4　图书 ISBN 校验器设计步骤

1. ISBN 编码的输入

利用 Scanner 类提供的 nextLine()方法可以实现字符串的输入。

3.4

2. digitAt()方法

digitAt()方法负责解析ISBN编码的每一位字符,并将其转换为对应的数字。首先利用字符串String类提供的char charAt(int index)方法得到字符串的第index个字符,然后将其转换为数字。

```java
public static int digitAt(String isbn, int index) {
    char digit = isbn.charAt(index);
    if(digit == 'X')
        return 10;
    else
        return digit - '0';
}
```

3. isISBN()方法

isISBN()方法判断读入的编码是否是合法的ISBN编码。

```java
public static boolean isISBN(String isbn) {
    boolean result = false;
    int checkSum = 0;
    if (isbn.length() == 10) {
        for (int i= 1; i< 10; i++ )
            checkSum += (11 - i) * digitAt(isbn, i - 1);
        checkSum = 11 - checkSum % 11;
        if (checkSum == digitAt(isbn, 9))
            result = true;
    } else if (isbn.length() == 13) {
        for (int i = 1; i < 13; i++ ) {
            if (i % 2 == 0)
                checkSum += 3 * digitAt(isbn, i - 1);
            else
                checkSum += digitAt(isbn, i - 1);
        }
        int remainder = checkSum % 10;
        if(remainder == 0)
            checkSum = 0;
        else
            checkSum = 10 - remainder;
        if (checkSum == digitAt(isbn, 12))
            result = true;
    }
    return result;
}
```

4. 定义主方法

定义主方法,调用上述方法,构建图书 ISBN 校验器。

```
public static void main(String[] args) {
    System.out.print("请输入图书的 ISBN 编码");
    Scanner sc = new Scanner(System.in);
    String isbn = sc.nextLine();
    if (isISBN(isbn))
        System.out.println("ISBN 校验码正确");
    else
        System.out.println("ISBN 校验码不正确");
    sc.close();
}
```

3.5 练一练

1. 按照如下个人所得税税率表及公式编写个人所得税计算器。

$$全月应纳税所得额 = 扣除四金之后的应发工资 - 5000$$

提示:个人所得税的计算可以采用以下两种方法。

(1) 超额累积税率计算法如表 3-1 所示。

表 3-1 超额累积税率计算法

级数	全月应纳税所得额	税率/%
1	不超过 3000 元	3
2	超过 3000 元至 12000 元的部分	10
3	超过 12000 元至 25000 元的部分	20
4	超过 25000 元至 35000 元的部分	25
5	超过 35000 元至 55000 元的部分	30
6	超过 55000 元至 80000 元的部分	35
7	超过 80000 元的部分	45

(2) 速算扣除数法计算法如表 3-2 所示。

工资个税的计算公式为

$$应纳税额 = 全月应纳税所得额 \times 适用税率 - 速算扣除数$$

表 3-2 速算扣除数法计算法

级数	全月应纳税所得额	税率/%	速算扣除数
1	不超过 3000 元	3	0
2	3000 元至 12000 元	10	2520

续表

级数	全月应纳税所得额	税率/%	速算扣除数
3	12000 元至 25000 元	20	16920
4	25000 元至 35000 元	25	31920
5	35000 元至 55000 元	30	52920
6	55000 元至 80000 元	35	85920
7	超过 80000 元	45	181920

2. 通过键盘输入任意多个数字并计算其平均值。

提示：文本扫描器类 Scanner 的 hasNextDouble() 可以判断是否存在下一个 double 类型的输入数据。

3. 用 $\dfrac{\Pi}{4} \approx 1 - \dfrac{1}{3} + \dfrac{1}{5} - \dfrac{1}{7} + \cdots$ 求 Π 的近似值，直到发现某一项的绝对值小于 10^{-6} 为止。

提示：对于"/"运算符，当两个操作数都为 int、long、short 时，计算结果只保留整数部分，舍弃小数部分。

第 4 章 数组——迷宫游戏

4.1 迷宫游戏介绍

设计实现一个简单的迷宫游戏。

(1) 在字符界面随机生成迷宫地图,如图 4-1 所示。

其中,1 表示障碍物,0 表示通路,2 表示奖励,8 表示当前位置。

游戏默认迷宫四周由障碍物围绕,入口位于地图的左上角,出口位于地图右下角。

(2) 用户在控制台输入"awsd"字符,分别代表方向键"←↑↓→";输入"x"字符,表示提前退出游戏。

(3) 在有限步数下,判断游戏的输赢,输出用户得分。

(4) 鼓励用户以较短路径、获得较多奖励完成游戏,设定如下积分原则:

① 玩家遇到奖励,积分增加奖励分值;

② 游戏结束时,积分增加游戏限定步长与当前游戏步长的差值。如限定步长为 30,玩家本次到达终点用了 20 步,则积分增加 10。

```
1111111111
1820201011
1220200101 1
1121122222 1
1201111021 1
1110112222 1
1010100111
1202022120 1
1020001121 1
1102002120 1
1111111111
```

图 4-1 迷宫地图

4.2 程序设计思路

1. 随机生成迷宫地图

游戏中需要随机生成迷宫地图,引入 java.util.Random 类,利用 Random 类提供的生成随机数方法随机生成障碍物、通路或者奖励等状态。迷宫地图采用二维数组进行表示与存储。

2. 判断玩家闯关是否成功

根据玩家的当前位置是否位于地图的出口判断玩家是否闯关成功。如果成功则根据游戏步数修改本次游戏的积分。

3. 游戏主控模块

游戏的主控模块是一个基于用户输入的循环控制模块,循环控制的条件是判断游戏是否结束。若游戏结束,则退出游戏主控模块,输出玩家的输赢状态及游戏积分;否则主控模块根据用户输入的方向键不断修改地图状态及更新玩家积分。

4.3 关键技术

4.3.1 什么是数组

数组是 Java 语言中常见的一种数据结构,用于存储相同类型的若干个变量。数组中存储的变量又称数组元素,数组元素可以通过数组索引访问。数组中能够保存的变量最大个数称为数组的长度。Java 语言中,数组的长度是固定的,在初始化时被确定下来。值得注意的是,数组也是一种数据类型,只不过与 Java 基本数据类型不同的是,数组是一种引用类型,也就是用数组变量指示数组的引用。

4.3.2 数组的定义及初始化

1. 定义

定义一维数组有下列两种格式:

数据类型[] 数组变量;
数据类型 数组变量[];

注意:定义数组时不能指定数组的长度。

数组是一种引用数据类型,当定义一个数组时,实际上是定义了一个引用变量,但是该引用变量还没有确切的内存地址与之相对应,下一步需要对数组完成初始化,才能真正地使用数组。

2. 初始化

Java 数组可以采用以下两种方式实现初始化。

方式 1:

数组变量 =new 数据类型[数组长度];

方式 1 中指定数组的长度,并且为每个数组元素指定初始值。初始化时的数据类型必须与定义的数据类型相同,或者是其子类类型。

系统按照表 4-1 所示的原则为数组元素分配初始值。

表 4-1 数组元素的初始化值

数据类型	byte、short、int、long	float、double	char	boolean	类、接口等引用类型
初始化值	0	0.0	'\u0000'	false	null

例如:

int[] arr;

```
arr=new int[10];
```

定义一个名为 arr 的 int 数组类型的变量,初始化时为数组变量 arr 分配可以容纳 10 个整数的内存空间。上述定义与初始化语句等价于 int[] arr = new int[10];。

方式 2:

数组变量 = new 数据类型[]{值 1, 值 2, 值 3, …};

或者将数组的定义与初始化合并为

数据类型[] 数组变量 ={值 1, 值 2, 值 3, …};

方式 2 中显式地指定每个数组元素的初始值,由系统根据数组元素的个数确定数组的长度。

4.3.3 数组的使用

1. 访问数组元素

数组变量[元素的索引值]

Java 语言中数组元素的索引值从 0 开始。

2. 数组的 length 属性

length 属性用于获取数组中数组元素的个数。

下面的例 4-1 示范了数组的定义与使用。

例 4-1

```
1.   public class DefArray {
2.       public static void main(String[] args) {
3.           int[] arr1 = new int[3];
4.           arr1[0] = 11;
5.           arr1[1] = 22;
6.           arr1[2] = 33;
7.           int[] arr2 = new int[5];
8.           arr2 = arr1;
9.           arr2[1] = 44;
             System.out.println("arr2 数组的长度为: " + arr2.length);    //输出 3
10.      }
11.  }
```

数组 arr1 与 arr2 的存储示意图如图 4-2 所示。程序在代码行 3 实现数组 arr1 的定义及初始化操作,arr1 的内存表示如图 4-2(a)所示;经过代码 4~6 行之后,对数组 arr1 的元素进行赋值,内存状态如图 4-2(b)所示;代码行 7 定义并初始化数组 arr2,如图 4-2 (c)所示;代码行 8 对数组变量 arr2 进行引用赋值,这时,数组变量 arr1 与 arr2 拥有同一个数组空间,如图 4-2(d)所示;其中没有被引用的数组对象就会变成"垃圾",将被垃圾回收机制回收。代码行 9 通过数组变量 arr2 修改数组元素的值,同样,数组元素 arr1[1]的值也会随之改变,内存状态如图 4-2(e)所示。

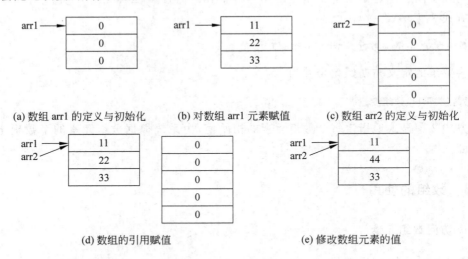

图 4-2 数组变量的存储示意图

4.3.4 数组的遍历

1. 方法 1

利用循环结构通过数组索引值的迭代实现数组的遍历。

```
for(循环变量=0;循环变量<数组变量.length;循环变量++){
    通过循环变量的迭代,遍历数组元素:数组变量[循环变量]
}
```

2. 方法 2

Java 5 之后,Java 提供了 foreach 循环语句,定义如下。

```
for(数据类型 变量:数组变量|集合){
    变量迭代访问每一个数组元素;
}
```

foreach 语句用于遍历数组或者集合(关于集合的介绍请参考本书第 9 章)元素。下面的代码段示范了数组的遍历。

```
for(int data:arr){
    System.out.println(data);
}
```

foreach 语句与普通循环的遍历方式不同,它无须数组索引值的迭代,而是利用临时变量(如上例的 data)自动访问数组的每个元素。

注意:在 foreach 语句中,通常不需要对循环变量进行赋值。

4.3.5 在方法中使用数组

Java 的方法是指一段完成特定功能的代码块,方法被封装为一个过程体,可以被重复调用,类似于 C 语言中的函数。方法的定义格式如下:

修饰符 返回值类型 方法名 (参数列表) {
    函数体;
    return 返回值;
}

方法通过参数与返回值实现方法调用过程信息的传递。其中,方法的参数与返回值都可以定义为数组变量,那么采用数组作为方法的参数和返回值,与采用普通变量进行信息传递有什么不同呢?

1. 数组作为方法的参数

与普通变量作为方法参数不同,数组作为方法参数传递的是数组的引用。所以,如果在方法中修改数组的元素值,则方法调用结束后,主调方法将得到修改之后的数组。

2. 数组作为方法的返回值

与数组作为方法参数相同,数组变量作为方法的返回值同样也返回数组的引用。

下面的例 4-2 示范了数组与方法的相互作用。

例 4-2

```java
public class ArrayMethod {
    public static void main(String[] args) {
        int a[] = { 1, 2, 3, 4, 5, 6, 7, 8, 9, 10 };
        reverse(a);
        for (int i : a)
            System.out.println(i);
        int b[] = copyArray(a, 2, 5);
        for (int i: b)
            System.out.println(i);
    }
    public static void reverse(int[] arr) {
        for (int i = 0, j = arr.length - 1; i< j; i++, j--) {
            int temp;
            temp = arr[i];
            arr[i] = arr[j];
```

```
            arr[j] = temp;
        }
    }
    public static int[] copyArray(int[] arr, int start, int length) {
        int[] result = new int[length];
        for (int i = start, j = 0; j < length; i++, j++) {
            result[j] = arr[i];
        }
        return result;
    }
}
```

3. 可变长度参数列表

Java 5 之后,Java 支持方法的可变长度参数列表,方法定义如下。

修饰符 返回值类型 方法名 (数据类型…参数变量) {
    …
}

在方法体中,对可变长度参数变量的使用等同于数组的操作。例如下面定义的方法可以输出所有的参数值。

```
public static void printParams(int … args) {
    for(int var:args) {
        System.out.println(var);
    }
}
```

调用该方法时,可以给出任意个参数,也可以以数组为实参。例如下面的调用方式都是合法的。

```
printParams();
printParams(1);
printParams(1,2);
printParams(new int[] {1,2,3,4,5});
```

注意:如果方法具有多个参数,则可变长度参数必须是方法的最后一个参数。

4.3.6 多维数组

1. 二维数组定义

数据类型[][] 数组变量;

2. 二维数组初始化

方式 1(等长二维数组):同时指定二维的长度。

数组变量 =new 数据类型[数组长度 1][数组长度 2];

例如:

```
int[][] arr;
arr=new int[2][3];
```

定义并初始化 2 行 3 列的二维 int 类型数组 a,数组 a 中的每个元素初始化为 0。

方式 2(不等长二维数组):先指定一维的长度,然后分别定义不同的二维长度。

例如:

```
int[][] arr=int[2][];
arr[0]=new int[3];
arr[1]=new int[4];
```

定义一个 2 行的二维数组 arr,数组 arr 变量由两个长度分别为 3 和 4 的一维数组 arr[0] 与 arr[1]组成。二维数组本质上依然是一维数组,其元素是指向一维数组的引用,arr 数组的内存示意如图 4-3 所示。

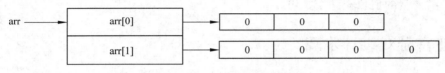

图 4-3 数组 arr 的内存示意图

4.3.7 Arrays 类

java.util.Arrays 类提供了许多可以直接操作数组的静态方法。

1. void sort(数据类型[] 数组变量)

对数组元素根据其自然顺序的升序进行排序。例如:

```
int[] arr = {56,43,78,23,58,456,78};
Arrays.sort(arr);          //数组 arr 经过排序之后的顺序:23 43 56 58 78 78 456
```

2. int BinarySearch(数据类型 数组变量,数据类型 关键字)

使用二分法在数组中查找指定关键字,如果查找成功,则返回关键字所在的索引,否则返回负数。

注意:调用该方法之前要保证数组按升序排列。例如,在上述有序数组 arr 中查找元素 58:

```
Arrays.binarySearch(arr, 58);          //返回值为 3
```

3. 数据类型[ ] copyOf(数据类型[ ] 原数组, int 长度)

复制原数组返回至新数组。其中,参数长度是新数组的长度。如果参数长度大于原数组的长度,则复制原数组的所有元素,后面的元素以 0(数组类型为数值型)或者 false(布尔类型)或者 null(引用类型)进行填充;如果参数长度小于原数组的长度,则截取原数组前面的部分元素。例如:

```
int[] arrCopy = Arrays.copyOf(arr, arr.length+1);
//复制生成数组 arrCopy,元素为: 56 43 78 23 58 456 78 0
```

4. boolean equals(数据类型 数组变量 1, 数据类型 数组变量 2)

判断两个数组是否相等。只有当两个数组包含相同数量的元素,并且所有对应位置的数组元素也相等时,才判定两个数组相等,方法返回 true,否则返回 false。

5. void fill(数据类型[ ] 数组变量, 数据类型 变量)

用参数变量的值填充整个数组。

4.4 迷宫游戏设计步骤

1. 定义符号常量

为提高程序的可读性,定义如下符号常量。

```java
public final static int ICON_NUMBER = 3;        //图标个数
public final static int WALL = 1;               //障碍物
public final static int ROAD = 0;               //通路
public final static int AWARD = 2;              //奖励
public final static int MAP_SIZE = 11;          //地图规模
public final static char UP = 'w';              //用户控制键
public final static char DOWN = 's';
public final static char LEFT = 'a';
public final static char RIGHT = 'd';
public final static char EXIT = 'x';
public final static int CUR_STATE = 8;          //标识当前位置图例
public final static int MAXSTEPS = 30;          //游戏的最大步数
static int curX = 0;                            //当前位置(curX,curY)
static int curY = 0;
static int score = 0;                           //游戏得分
static int steps = 0;                           //步数
```

2. 随机生成地图

游戏默认迷宫四周由障碍物围绕,入口位于地图的左上角,出口位于地图右下角。利用 Java 随机数类 java.util.Random 提供的 nextInt(int bound) 方法实现区间[0,bound)内伪随机数的生成。例如,"Random random = new Random(); random.nextInt(3);"随机生成[0,2]区间内的整数。

```java
public static void generateMapRandom() {
    Random random = new Random();
    for(int line = 0; line < map.length; line++) {
        map[line][0] = Maze.WALL;
        map[line][map.length - 1] = Maze.WALL;
        for(int col = 1; col < map[line].length-1; col++) {
            if(line == 0 || line == map.length-1)
                map[line][col] = Maze.WALL;
            else
                map[line][col] = random.nextInt(Maze.ICON_NUMBER);
        }
    }
    map[1][1] = Maze.CUR_STATE;          //设置迷宫的入口及出口
    curX = 1;
    curY = 1;
    map[map.length-2][map.length-2] = Maze.ROAD;
}
```

3. 绘制地图

遍历并打印二维迷宫数组。

```java
public static void printMap() {
    for (int line = 0; line < map.length; line++) {
        for (int col = 0; col <= map[line].length - 1; col++)
            System.out.print(map[line][col] + " ");
        System.out.println("");
    }
}
```

4. 获取用户输入

根据游戏要求,单字符输入可以使用 System.in 的 read()方法,将输入的字节数据转换为字符。利用循环语句,仅接收合法的方向按键。

```java
public static char readDirectionKey() throws IOException {
    char direction = (char)System.in.read();
    while(direction != Maze.LEFT&&direction != Maze.RIGHT
            &&direction != Maze.UP&&direction != Maze.DOWN
            && direction != Maze.EXIT)
        direction = (char)System.in.read();
    return direction;
}
```

5. 判断玩家是否闯关成功

根据玩家的当前位置是否位于地图的出口来判断玩家是否闯关成功。如果成功,则根据游戏步数修改本次游戏积分。

```
public static boolean isSuccess() {
    if (map[map.length - 2][map.length - 2] == Maze.CUR_STATE) {
        score += Maze.MAXSTEPS - steps;
        return true;
    } else
        return false;
}
```

6. 游戏主控程序

playGame()方法首先判断游戏是否结束,若未结束,则读取用户输入的控制键,根据用户输入的方向键修改地图状态、更新玩家积分;若游戏满足结束条件,则判断玩家是否闯关成功,如果闯关成功,则playGame()方法返回true;否则返回false。

```
public static boolean playGame() {
    int state = Maze.ROAD;
    while (!Maze.isSuccess() &&steps < Maze.MAXSTEPS) {
        map[curX][curY] = state;           //恢复地图原有的图标
        try {
            switch (readDirectionKey()) {
            case Maze.UP:
                if (map[curX - 1][curY] != Maze.WALL) {
                    score += map[curX - 1][curY];
                    state = map[curX - 1][curY];
                    map[curX - 1][curY] = Maze.CUR_STATE;
                    curX--;
                    steps++;
                }   break;
            case Maze.DOWN:
                if (map[curX + 1][curY] != Maze.WALL) {
                    score += map[curX + 1][curY];
                    state = map[curX + 1][curY];
                    map[curX + 1][curY] = Maze.CUR_STATE;
                    curX++;
                    steps++;
                }   break;
            case Maze.LEFT:
                if (map[curX][curY - 1] != Maze.WALL) {
```

```
                    score += map[curX][curY - 1];
                    state = map[curX][curY - 1];
                    map[curX][curY - 1] = Maze.CUR_STATE;
                    curY--;
                    steps++;
                }   break;
            case Maze.RIGHT:
                if (map[curX][curY + 1] != Maze.WALL) {
                    score += map[curX][curY + 1];
                    state = map[curX][curY + 1];
                    map[curX][curY + 1] = Maze.CUR_STATE;
                    curY++;
                    steps++;
                }   break;
            }
            case Maze.EXIT:
                System.out.println("放弃挑战,退出游戏");
                System.exit(- 1);
            }
        } catch (IOException e) {
            e.printStackTrace();
        }
        Maze.printMap();
    }
    return isSuccess();
}
```

7. 主程序

```
public static void main(String[] args) {
    Maze.generateMapRandom();
    Maze.printMap();
    System.out.println("请输入 a(左) s(下) w(上) d(右) x(退出)");
    if(Maze.playGame())
        System.out.println("恭喜你,闯关成功!!! 得分: "+ score);
    else
        System.out.println("很遗憾,挑战失败,加油哦!!");
}
```

4.5 练一练

1. 程序填空:根据程序注释完成下列两个方法的定义。

```
public class BubbleSort {
```

```java
public static void sort(int[] data) {        //采用冒泡法对参数数组进行排序
    ------------------------------------------------------------
}
//返回 data 数组中小于 reference 的所有值
public static int[] smaller(int reference, int[] data) {
    ------------------------------------------------------------
}
public static void main(String[] args) {
    int[] array = { 34, 56, 45, 13, 78, 56, 33 };
    System.out.println("排序数组元素为：");
    for (int i = 0; i < array.length; i++)
        System.out.print(array[i] + " ");
    sort(array);
    System.out.println("\n 排序后的数组为：");
    for (int num: array) {
        System.out.print(num + " ");
    }
    System.out.println("\n 小于 50 的数组元素为：");
    for(int num: smaller(50,array)) {
        System.out.print(num + " ");
    }
}
}
```

2. 试分析下列代码出现错误的原因。

```java
package chap04;
import java.util.Arrays;
public class VariablePara {
    public static int[] smaller(int...data) {
        int[] temp = new int[data.length];
        int count = 0;
        for(int i = 1; i < data.length; i++)
            if(data[0] > data[i])
                temp[count++] = data[i];
        return Arrays.copyOf(temp, count);
    }
    public static int[] smaller(int reference, int... data) {
        int[] temp = new int[data.length];
        int count = 0;
        for(int i = 0; i < data.length; i++)
            if(reference> data[i])
                temp[count++] = data[i];
        return Arrays.copyOf(temp, count);
    }
```

```java
public static void main(String[] args) {
    System.out.println("\n 小于 60 的元素为: ");
    for(int num: smaller(60,new int[] {55,66,77,33,44})) {
        System.out.print(num + " ");
    }
    for(int num: smaller(60,55,66,77,33,44)) {
        System.out.print(num + " ");
    }
}
```

Eclipse 中出现语法错误，提示如下：

```
public static void main(String[] args) {
    System.out.println("\n小于60的元素为：");
    for(int num : smaller(60,new int[] {55,66,77,33,44})) {
        System.out.print(num + " ");
    }
    for(int num : smaller(60,55,66,77,33,44)) {
        System.out
```
The method smaller(int[]) is ambiguous for the type VariablePara
Press 'F2' for focus

3. 利用 Arrays 类的常用方法完成下列源程序。

```java
import java.util.Arrays;
public class ArraysTest {
    public static void main(String[] args) {
        int[] array1 = new int[] {22,33,55,11,66,77,88,44,99};
        //复制生成 array1 的副本 array2
        _____
        //对数组 array1 进行升序排列
        _____
        System.out.println("对数组 array1 进行排序,结果为: "
                +Arrays.toString(array1));
        //判断数组 array1 与 array2 是否相等,并输出结果
        _____
        //采用二分查找方法,在数组 array1 中查找 55,返回位置赋值 location 变量
        _____
        if(location >= 0)
            System.out.println("查找数组元素 55 下标为: "+location);
        else
            System.out.println("数组中不存在元素 55");
    }
}
```

第 5 章 类与类的继承——个人通讯录（一）

5.1 个人通讯录（一）

设计简单的个人通讯录系统（一），用于存储联系人信息，实现对联系人按照基本联系人、家人、工作伙伴等进行分类存储，具体要求如下：
（1）联系人的基本信息包括姓名、性别、电子邮件及若干个联系电话；
（2）家人除具有联系人基本信息之外，还需要定义家庭地址及生日；
（3）工作伙伴除具有联系人基本信息之外，还需要添加所在公司及职称/职务信息。

5.2 程序设计思路

实体对象可以看作是具有内部属性及功能的构件，通过封装技术隐藏实体内部的信息及功能的实现细节。个人通讯录（一）涉及 3 个基本类，分别是联系人类、家人类及工作伙伴类。

联系人类封装联系人的基本信息及输出功能；家人类复用联系人类并添加特有属性，描述联系人的家庭地址与生日；工作伙伴类也复用联系人类，同时通过组合技术描述工作伙伴所在的公司及职称/职务信息。

5.3 关键技术

5.3.1 面向对象的基本概念

Java 是面向对象的程序设计语言，其基本思想是通过对象、类、封装、继承、多态等基本概念进行程序设计，它是从现实世界中存在的客观事物（即对象）出发构建软件系统的。

对象：作为构成系统的基本单位，对象是由数据及其行为所构成的封装体。对象包含 3 个基本要素，分别是对象标识、对象状态和对象行为。每一个对象必须有一个名字，以区别于其他对象，这就是对象标识；状态用来描述对象的某些特征；对象行为用来封装对象所拥有的业务操作。

例如，联系人 Contract 对象包含姓名、性别、电子邮件及联系电话等基本状态信息，同时还具有打印输出的行为特征。

类：对同类对象进行抽象形成类，它为属于该类的所有对象提供了统一的抽象描述，其内部包括状态和行为两个主要部分。类也可以被认为是一种自定义的数据类型，可以使用类定义变量，所有通过类定义的变量都是引用变量，它们将会引用到类的实例，即对象。

封装：封装是一种信息隐蔽技术，通过封装将对象的状态和行为结合成一个独立的模块，尽可能隐藏对象的内部细节（包括对象内部的私有状态及行为的具体实现）。封装的目的在于把对象的设计者与使用者分开，作为使用者不必了解对象内部的实现细节，只需要使用设计者提供的行为方法实现功能即可。

继承：继承表示类之间的层次关系。继承关系使得子类对象可以共享父类对象的状态和行为。继承又可分为单继承（一个子类仅拥有一个父类）和多继承（一个子类可以拥有多个父类），Java语言支持类的单继承，而C++允许多继承。在程序设计过程中通过继承性，一方面得到了类的层次等级结构；另一方面，通过类的继承关系可以使公共的状态和行为特性得到共享，提高了软件的重用性。

多态：多态性是指同名的行为方法可在不同的类中具有不同的实现。在子类继承父类的同时，类的方法实现可以进行扩充或者修改，使子类的同名方法更适合子类对象。如父类图形（Shape）的绘图方法draw()，在其子类圆形（Circle）和正方形（Square）都具有同名的绘制方法draw()，但是绘制的内容和方式都是不同的。

5.3.2 类的定义

类是一种自定义的引用型数据类型，可以采用如下格式实现类的定义：

```
[public|abstract|final] class 类名{
    [初始化块的定义;]
    [成员变量的定义;]
    [方法的定义;]
}
```

public、abstract、final称为修饰符，分别用于定义类的访问权限、抽象类及最终类等属性。方括号（[]）表示可选项。类名必须是合法的Java标识符。为了提高程序的可读性，Java的类名通常由若干个有意义的单词组成，每个单词首字母大写，其余字母全部小写。

大括号（{}）之间的内容称为类体，主要包括3部分：初始化块、成员变量的定义、方法的定义。类中各成员之间的定义顺序没有任何影响，各成员之间可以互相调用。初始化块用于对类对象进行初始化操作，详细内容见5.3.8节；成员变量用于定义同类对象具有的状态数据；方法则定义类的行为特征或功能实现。

1. 成员变量

成员变量通常用于描述类或者类实例的状态信息，其定义格式如下：

[修饰符] 数据类型 成员变量[=默认值];

成员变量的修饰符用于定义成员变量的属性，如public、protected、private、static、final等。成员变量的数据类型既可以为基本数据类型，也可以是引用数据类型，如数组、类及接口等。

2. 方法

方法通常描述类或者类实例的行为及功能，其定义格式如下：

[修饰符] 返回值类型 方法名(形参列表) {
//方法体
}

修饰符用于定义成员方法的属性，如 public、protected、private、static、final 等。方法的返回值类型既可以为基本数据类型，也可以是引用数据类型。

下面的例 5-1 示范了 Student 类的定义。

例 5-1

```
public class Student {
    String id;                              //学号
    String name;                            //姓名
    Date birth;                             //生日
    float[] score;                          //成绩
    public float computeAverage() {         //计算平均成绩
        float sum = 0.f;
        for(float s:score)
            sum+=s;
        return sum/score.length;
    }
    public String toString() {              //对象的字符串表示
        return id+"\t"+name;
    }
}
```

类 Student 用于描述学生信息，包括 4 个成员变量，其中，成员变量 birth、score 是引用类型变量；定义了两个公有（public）方法：public float computeAverage()和 public String toString()。

3. 成员变量与局部变量

在 Java 语言中，根据定义变量位置的不同，可以将变量分成两大类：成员变量与局部变量。成员变量、局部变量的命名必须遵守 Java 语言关于标识符的命名规则，通常由多个有意义的单词组成，首单词小写，后面每个单词首字母大写。

局部变量根据定义形式的不同，又可以分为形参变量（方法的形参变量）、方法局部变量（在方法体内定义的局部变量）及代码块局部变量（代码块中定义的局部变量）。

成员变量与局部变量在 Java 程序中存在如下差异。

(1) 在类中的位置不同。

成员变量：定义位于类体中，方法的外面；

局部变量：在方法或者代码块中，或者方法的声明上（即在参数列表中）。

(2) 生命周期不同。

成员变量：随着对象的创建而存在，随着对象的消失而消失，其作用域为类中的所有方法。

局部变量：随着方法的调用或者代码块的执行而存在，随着方法的调用完毕或者代码块的执行完毕而消失，局部变量在其所定义的方法或者代码块中有效。

（3）初始值。

成员变量：无须显式初始化，有默认初始值，如表 5-1 所示。

表 5-1　成员变量默认的初始值

数据类型	byte、short、int、long	float、double	char	boolean	类、接口等引用类型
初始化值	0	0.0	'\u0000'	false	null

局部变量：没有默认初始值，使用之前必须初始化。

Student 类的 id、name、score、birth 都是成员变量，可以被类的所有成员方法访问；成员方法 computeAverage() 中定义的变量 sum、s 是局部变量，sum 在 computeAverage() 的方法体中有效，局部变量 s 仅在 foreach 循环语句中可以被访问。局部变量没有初始值，需要先初始化，然后才能使用。

注意：Java 语言允许局部变量与成员变量同名。如果局部变量与成员变量同名，那么方法内的变量名默认表示局部变量；如果需要在方法中访问被隐藏的成员变量，则需要使用 this 关键字。

例如：

```
public void setName(String name) {
    this.name = name;
}
```

方法 setName() 中赋值运算的左侧通过 this 关键字访问成员变量 name；而赋值运算的右操作数 name 则是同名的局部变量（形参变量）。

5.3.3　对象

对象是类的实例，拥有具体的状态与行为。例如 Student 是类，描述同类对象共同的状态及行为，而对象是指一个具体的 Student 个体，例如姓名是"李白"的 Student 对象。对象是动态的，每个对象都拥有一个从创建、运行到消亡的动态过程。同时，对象也要占用内存空间，存储对象的状态数据，即成员变量。

1. 声明对象

对象必须先声明再使用，声明对象的一般格式为：

类名 对象名；

例如："Student stu;"通过声明定义对象变量 stu，表明该变量可以指向一个 Student 类型的对象。

2. 创建对象

创建对象就是通过 new 运算符调用类的构造方法（构造方法的详细内容见 5.3.5 节），创建类的实例对象。一般格式为：

对象名 = new 类名([参数列表]);

例如："stu = new Student();"创建了一个学生对象，并将其引用赋值给变量 stu。可以将对象的声明与创建合并成语句，如"Student stu = new Student();"。

3. 使用对象

创建对象之后，就可以通过以下两种方式使用对象。

（1）访问对象的成员变量。

对象变量.成员变量

（2）调用对象的方法。

对象变量.成员方法([参数列表])

4. 对象的运算

Java 语言支持对象的"=="与"!="关系运算，用于比较两个对象变量是否指向同一内存地址。

5. 对象的内存模型

本节分析对象在创建过程中的内存模型。上述代码行"Student stu = new Student();"在运行的过程中不仅生成一个 Student 对象，而且会同时产生一个 Student 类型的变量 stu。那么生成的对象和变量 stu 在内存中是如何存储的呢？

Java 程序在运行时，需要在内存中分配空间。为了提高运算效率，Java 程序对内存空间进行了不同的划分，其中就包括堆内存与栈内存。栈内存主要存放一些基本类型的变量（int、short、long、byte、float、double、boolean、char）和对象变量；而堆内存则用来存储 Java 中的对象。

（1）创建一个对象的内存模型。

代码行：

```
Student stu1 = new Student();
stu1.name = "李白";
stu1.id = "17010";
```

在内存中的存储示意图如图 5-1 所示。

图 5-1 指示了栈内存与堆内存空间。通过运算符 new 实例化 Student 对象，系统在堆内存中分配一段空间，用于存储 Student 对象的多个成员变量。同时，声明的对象变量 stu1 被存储在栈内存中，它指向刚刚创建的存储在堆内存中的 Student 对象，也就是变量

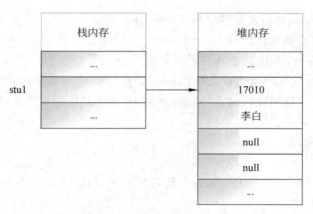

图 5-1 一个对象的内存模型

stu1 中存储了 Student 对象的引用。

（2）创建多个对象的内存模型

代码行：

```
Student stu1 = new Student();
stu1.name = "李白";
stu1.id = "17010";
Student stu2 = new Student();
stu2.name = "李白";
stu2.id = "17010";
```

在内存中的存储示意图如图 5-2 所示。

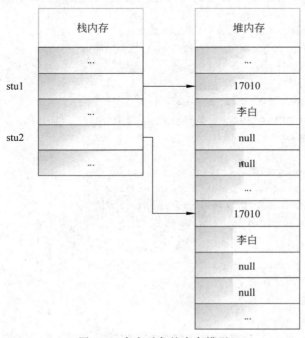

图 5-2 多个对象的内存模型

通过运算符 new 先后在堆内存中实例化两个 Student 对象。与对象相对应的对象变量 stu1 与 stu2 被存储在栈内存中。虽然堆内存保存的两个 Student 对象属性相同,但是其位于不同的内存空间,也就是具有不同的引用。

若执行代码行"stu1＝null;",这时,stu1 变量指向的存放于堆内存中的对象就变成了"垃圾",Java 的垃圾回收机制将回收该对象,释放该对象所占用的堆内存空间。

5.3.4 方法的重载

Java 语言允许在一个类中定义多个同名的方法,但是方法的参数必须不同。同名的方法被称为重载方法。

重载方法通过不同的参数个数或参数类型进行区别,返回值类型及方法的修饰符定义都与重载无关。

例 5-2 示范了方法的重载。

例 5-2

```java
public class Calculator{
    public int add(int a, int b) {
        System.out.println("参数是 int 类型");
        return a+b;
    }
    public float add(float a, float b) {
        System.out.println("参数是 float 类型");
        return a+b;
    }
    public double add(double a, double b) {
        System.out.println("参数是 double 类型");
        return a+b;
    }
    public byte add(byte a, byte b) {
        System.out.println("参数是 byte 类型");
        return (byte)(a+b);
    }
    public staticvoid main(String[] args) {
        Calculator cal = new Calculator();
        //传递整数参数,所以调用 add(int,int)方法
        System.out.println(cal.add(1, 3));
    }
}
```

5.3.5 构造方法

在类的定义中,有一类特殊的方法称为构造方法。构造方法与类同名,没有返回值,

可以通过构造方法实现实例对象的初始化操作。

1. 构造方法的定义

一般格式如下：

```
class 类名{
    public 构造方法名([参数列表]){
        方法体
    }}
```

2. 构造方法的调用

new 运算符将自动调用构造方法，实现对象实例的初始化操作。例如：

```
public class Student {
    String id;                //学号
    String name;              //姓名
    public Student(String strId, String strName) {    //定义构造方法
        id = strId;
        name = strName;
    }
    public static void main(String[] args) {
        Student stu = new Student("17010","李白");    //自动调用构造方法
        System.out.println(stu.name);                 //输出"李白"
    }
}
```

3. 默认的构造方法

如果一个类中没有自定义构造方法，则系统会自动生成一个无参数且方法体为空的构造方法，称为默认的构造方法；反之，如果类中已经显式定义了构造方法，则系统不会再自动生成默认的构造方法。

4. 构造方法的重载

构造方法也可以重载，即提供多种逻辑以实现类实例对象的初始化。例如：

```
public class Student {
    String id;                        //学号
    String name;                      //姓名
    public Student() {                //定义默认构造方法
    }
    public Student(String strId) {    //定义一个String参数的构造方法
        id = strId;
    }
```

```java
        public Student(String strId, String strName) {
                                        //定义两个String参数的构造方法
            id = strId;
            name = strName;
        }
        public static void main(String[] args) {
            Student stu1 = new Student();          //调用默认的构造方法创建Student对象
            System.out.println(stu1.name);
            Student stu2 = new Student("17010");
                                        //调用单参数的构造方法创建Student对象
            System.out.println(stu2.name);
            Student stu3 = new Student("17010","李白");
                                        //调用两个参数的构造方法创建对象
            System.out.println(stu3.name);
        }
    }
```

注意：Student类不能再定义。

```java
public Student(String strName) {
    name = strName;
}
```

编译时，系统提示"Duplicate method Student(String) in type Student"，即出现方法重复定义的错误。

5.3.6 this 关键字

Java语言提供了this关键字，代表当前对象的引用。this关键字有以下3种典型应用。

1. 在类的实例方法中使用this访问本类的其他成员

例如：

```java
public class Monkey {
    float height;              //身长
    float weight;              //体重
    Container container;       //装载猴子的笼子
    public String toString() {
        return "Monkey height:"+this.height+";weight:"+this.weight;
    }
    void print() {
        System.out.println("当前对象信息:"+this.toString());
    }
```

```
}
class Container{
    Monkey monkey;
    Container(Monkey monkey){
        this.monkey = monkey;
    }
}
```

上述类 Monkey 的实例方法 print 中使用了 this 关键字调用了另一个实例方法 toString()；在实例方法 toString()中，又通过关键字 this 访问了当前对象的 height 实例变量。通常情况下，以上两处可以省略"this."，默认指代当前对象的引用。但是，在下面构造方法中：

```
Monkey(float height, floatweight) {
        this.height = height;
        this.weight = weight;
}
```

由于实例变量(height)与局部变量同名，所以必须使用 this 关键字区别实例变量。

2. 在类的构造方法中使用 this 调用本类的另一个构造方法

通过 this 关键字调用构造方法只能用在构造方法中，且它必须是第一行语句。例如，在 Monkey 类中增加无参的构造方法：

```
Monkey() {
        this(50.0f,10.0f);
}
```

上述构造方法中使用 this(float,float)，实际上是调用另一个构造方法 Monkey(float,float)实现 Monkey 对象的初始化操作。

3. 在类的实例方法中 this 指代当前对象的引用

例如，在 Monkey 类中定义实例方法：

```
void loadContainer() {
    container= new Container(this);       //将当前 Monkey 对象装到 container 中
}
```

通过 this 关键字指代当前对象，将当前的 Monkey 对象装到 container 对象中。

5.3.7 static 关键字

在 Java 类的定义中，static 关键字可用于声明成员变量、方法与初始化块(关于初始化块的详细内容参见 5.3.8 节)。

1. 静态变量与实例变量

以 static 修饰的成员变量称为类变量或者静态变量,反之则称为实例变量。类变量属于类所有,描述类的状态信息,为类的所有对象共享。其生存周期开始于类的加载阶段,结束于类的销毁;实例变量属于类的实例对象所有,描述对象个体的状态信息,生存周期开始于实例对象的构造,结束于实例对象的销毁。

类变量的访问方式:

类名.类变量

实例变量的访问方式:

对象.实例变量

例 5-3 示范了静态变量与实例变量的使用。

例 5-3

```java
public class Employee {
    int id;                              //定义实例变量:员工编号
    String name;                         //定义实例变量:员工姓名
    static int nextId = 1;               //定义类变量:下一名待生成员工的编号
    public Employee() {
        id = Employee.nextId;
        Employee.nextId++;
    }
    public Employee(String strName) {
        this();
        name = strName;
    }
    public static void main(String[] args) {
        Employee ee1 = new Employee();
        Employee ee2 = new Employee("李白");
        //ee1.id = 1 ee1.name= null
        System.out.println("第一名员工:"+ee1.id+" "+ee1.name);
        //ee2.id = 2 ee2.name=李白
        System.out.println("第二名员工:"+ee2.id+" "+ee2.name);
        //通过对象变量 ee1、ee2 及类访问类变量 nextId,输出结果:3 3 3
        System.out.println("类变量:"+Employee.nextId+" "+ee1.nextId
                            +" "+ee2.nextId);
    }
}
```

2. 静态方法与实例方法

以 static 修饰的方法称为类方法或静态方法,反之则称为实例方法。在程序设计中要注意区别应用静态方法与实例方法：

（1）静态方法属于类所有，类实例化前即可使用，采用"类名.静态方法"的格式调用；

（2）实例方法可以访问类中的任何成员，静态方法只能访问类中的静态成员，并且 Java 语法规定，静态方法中不能使用 this 关键字。

5.3.8 初始化块

Java 实例化对象时要调用类的构造方法，实现对象的初始化。除构造方法外，类中的初始化块也可以对 Java 对象进行初始化。

1．定义初始化块

定义初始化块的语法规范如下：

```
[修饰符] class 类名{
    [static]{
    //可执行代码序列
    }
}
```

2．普通初始化块

如果代码块的定义中不包含 static 修饰符，则定义了一个普通的初始化块，它作为构造方法的补充一同参与对象的初始化工作。在创建实例对象时，系统首先调用普通初始化块，然后调用构造方法以共同完成对象的初始化工作。

3．静态初始化块

以 static 修饰的初始化块称为静态初始化块或类初始化块。普通初始化块与构造方法一同负责对象的初始化工作；静态初始化块则负责完成类的初始化工作，通常负责实现类变量的初始化。

静态初始化块仅在第一次加载该类时执行一次，并且如果类的直接或者间接父类也包含静态初始化块，则会先于当前类的静态初始化块被执行。普通初始化块在每次创建对象时都会被执行。

例 5-4 示范了初始化块的使用。

例 5-4

```
public class Employee {
    int id;
    String name;
    static int nextId;
    static{
        System.out.println("加载静态初始化块");
```

```
            Employee.nextId = 1;
        }
        {
            id = Employee.nextId;
            Employee.nextId++;
            System.out.println("初始化方法中实现"+id+"员工的初始化");
        }
        public Employee() {
            System.out.println("初始化方法中实现"+id+"员工的初始化");
        }
        public Employee(String strName) {
            name = strName;
            System.out.println("初始化方法中实现"+id+"员工的初始化");
        }
        public static void main(String[] args) {
            Employee ee1 = new Employee();
            Employee ee2 = new Employee("李白");
            System.out.println("第一名员工:"+ee1.id+" "+ee1.name);
            System.out.println("第二名员工:"+ee2.id+" "+ee2.name);
        }
    }
```

Java程序执行后,第一次遇到Employee时会加载Employee类,并执行静态代码块以实现类变量的初始化,后续两次实例化Employee对象都不再调用静态代码块,而是通过普通代码块和构造方法实现对象的初始化。

5.3.9 访问控制符

封装作为面向对象编程的重要特征之一,它可以将对象的状态信息隐藏在对象内部,不允许外部程序直接访问,让使用者只能通过类中事先定义的方法访问及修改状态信息。通过封装技术,一方面可以隐藏对象内部的状态信息及实现方法;另一方面可以暴露访问状态信息的方法,让使用者通过安全受控的方式访问状态信息。

为了实现类的封装性,要分别为类及类中的成员(变量及方法)定义不同的访问权限。

1. 类的访问权限

类的访问控制符包括public与缺省(default),类定义时,使用public修饰,称为公有类,可以被其他所有类访问;不使用任何修饰符定义的类具有缺省(default)访问权限,仅可被同一个包中的其他类使用,在其他包中该类不可见。

2. 类成员的访问权限

类成员的访问控制符包括public、protected、private与缺省(default),如表5-2所示。

表 5-2　成员的访问控制

	public	protected	default	private
同一个类中	√	√	√	√
同一个包中	√	√	√	
子类中	√	√		
全局范围的所有类	√			

（1）public：公共访问权限，被 public 修饰的成员拥有最大范围的访问权限，该成员可以被所有类对象访问。

（2）protected：子类访问权限，被 protected 定义的成员可以被同一类中的其他成员、同一包中的其他类或者不同包的子类访问。使用 protected 修饰符的重点在于设置该成员可以在其子类中被直接访问。

（3）缺省（default）：包访问权限，拥有包访问权限的成员可以被同一类的其他成员及同一包中的其他类访问，不能被其他包中的类访问。

（4）private：私有访问权限，被 private 定义的成员拥有最小范围的访问权限，仅能被同一类中的其他成员访问，将 private 成员隐藏在类的内部。

关于类成员的访问权限，存在如下一般原则：
- 类中的成员变量通常都定义为私有（private）访问权限，这体现了封装技术关于信息的隐蔽；
- 如果有子类需要用到父类的某些成员，则在父类中使用 protected 修饰这些成员，方便控制其可以在子类中被访问，而无法在不同包的其他类中被访问；
- 类中的成员方法一般定义为公有（public）访问权限，以便外部通过公有成员方法与该类发生联系。

5.3.10　包的定义及其导入

设计应用程序，尤其是多人合作开发规模较大的应用程序时，需要定义很多类实现程序功能，往往会涉及多个类重名的现象。为了解决类的同名冲突问题，Java 语言提供了管理类名空间的包。

1. 包的作用

（1）解决类的同名冲突

与 Windows 文件系统对文件的命名规则相同，Java 系统规定：同一个包中不允许有相同名字的类存在，不同包中的类名字可以相同。

（2）分类管理类文件

如同文件系统中的文件夹，包采用树状目录结构，可以将功能相近的类组织在同一个包中，便于分门别类地管理类文件。

(3) 限定包访问权限

对于缺省访问控制符的类及成员,默认具有包访问权限,也就是仅允许同一包中的类访问具有包访问权限的类及成员,而包之外的其他类都不可以访问。

2. 包的定义

包的定义规范如下：

```
package 包名;
```

包的定义必须是源程序中除注释语句之外的第一条语句。包名通常要小写,有一个或者多个有意义的单词,以"."间隔而成。

3. 包的导入

Java 程序中,如果使用了 java.lang 包之外的类,则必须先导入再使用。通过 import 关键字可以向当前的 Java 文件导入指定的类或者全部类。

```
import 包名.类名;
import 包名.*;
```

import 语句必须放置在所有的类定义之前和 package 语句之后。

5.3.11 类的继承

继承是面向对象程序设计的第二大特征,它是实现软件复用的重要手段,也是面向对象的基石。继承通过扩展一个已有类的实现得到新的类定义,形成父类与子类之间的继承关系。Java 语言仅支持单继承,即一个子类只能拥有一个父类。

1. 通过继承定义子类

Java 语言中通过继承关系定义子类的语法规范如下：

```
[public][abstract][final] class 类名 extends 父类名{
    成员的定义;
}
```

通过 extends 关键字,子类扩展自父类,形成父类与子类之间一般与特殊的关系。java.lang.Object 类是所有类的直接或者间接父类。

2. 继承原则

(1) 子类继承父类中所有的实例变量及类变量。

(2) 子类继承父类中除构造方法及初始化块之外的其他方法,包括实例方法及类方法。

(3) 子类可以增加新的方法及变量。

(4) 子类可以重写父类方法,但是不能删除父类方法或者变量。
(5) 子类对父类成员的访问权限如表 5-3 所示。

表 5-3 子类对父类成员的访问权限

父类成员的访问权限修饰符	public	protected	default	private
同一包中的子类对父类成员的访问	√	√	√	
不同包中的子类对父类成员的访问	√	√		

以下代码行通过继承关系由父类 Employee(例 5-4)定义了子类 Administrator。

```
public class Administrator extends Employee{
    private float overtimePay;              //子类中增加的实例变量与方法
    public float getOvertimePay() {
        return overtimePay;
    }
    public static void main(String[] args) {
        Administrator admin = new Administrator();
        admin.setName("张思明");              //子类 Administrator 继承父类 Employee
                                            //中定义的方法
        System.out.println(ee.getName());
    }
}
```

在 main()方法中增加下列代码:

`Administrator admin = new Administrator("张思明");`

这时会提示出错,这是因为子类 Administrator 无法继承父类的构造方法 Employee (String)。

5.3.12 子类重写父类方法

子类可以重新定义与父类相同的方法,实现对父类方法的覆盖,这样做需要注意以下问题。

(1) 子类方法名称、参数列表与父类方法相同。
(2) 子类方法的返回值:
- 若父类被重写方法的返回值类型是 void,则子类重写方法的返回值类型只能是 void。
- 若父类被重写方法的返回值类型是引用类型,如类 Object,则子类重写方法的返回值类型可以是类 Object 或 Object 类的子类,如 String 类型。
- 若父类被重写方法的返回值类型是基本数据类型,如 int,则子类重写方法的返回值类型必须是相同的基本数据类型。

(3) 子类方法抛出的异常类不能大于父类方法抛出的异常类(关于异常的详细内容请参见第 8 章)。

(4) 子类方法的访问权限要大于或等于父类方法的访问权限。

(5) 子类方法不能改变父类方法的 static 修饰。

5.3.13 super 关键字

Java 语言提供了 super 关键字,它指向当前对象的父类对象。super 关键字有以下两种典型应用。

1. 在子类构造方法中通过 super 显式调用父类的构造方法

子类继承父类中除构造方法之外的成员,所以在子类中需要定义构造方法从而实现子类对象的初始化。在子类的构造方法中,通常通过 super 关键字显式调用父类的构造方法,完成父类实例变量的初始化,然后实现子类中新添加实例变量的初始化操作。例如,在上述子类 Administrator 中增加构造方法的定义的方法如下:

```
public Administrator(String name, float overtimePay) {
    super(name);                        //调用父类构造方法 Employee(String)
    setOvertimePay(overtimePay);        //初始化新增的实例变量
}
```

注意:通过 super 调用的父类构造方法应该位于子类构造方法中的第一条语句。

2. 在子类实例方法中通过 super 关键字访问父类中被重写的成员

当子类重新定义了父类中的实例方法或者实例变量时,如果子类中需要访问被重写的实例变量或者实例方法,则需要通过关键字 super 指代父类中的同名成员。

注意:在子类类方法中不能使用 super 关键字访问父类的实例方法或实例变量。

5.3.14 final 关键字

final 关键字可以用来声明变量、方法及类。

1. final 成员变量

final 关键字可用于成员变量声明,它既可以修饰实例变量,也可以修饰类变量。final 成员变量表示初始化之后就不可以再改变该变量的值。通常,由 final 定义的变量为常量。通常,被定义为 final 的常量需要使用大写字母命名,如果名字包含多个单词,则中间使用下画线连接。

2. final 局部变量

final 关键字还可以修饰方法中的局部变量或者形参,表示只能对 final 变量执行一次赋值操作。

例 5-5 示范了 final 修饰局部变量及成员变量的使用。

例 5-5

```
public class FinalVar{
    public static final int var1 = 10;    //final 修饰类变量,在定义的同时赋值
    public static final int var2;         //final 修饰类变量,在初始化块中赋值
    static {
        Var2 = 20;
    }
    private final int var3;               //final 修饰实例变量,在构造方法中赋值
    public FinalVar() {
        var3 = 30;
    }
    public void function1(final int arg) {
        arg = 40;                         //final 修饰基本类型形参,赋值语句出错
        final int data;
        data = 50;                        //final 修饰的局部变量,第一次赋值正确
        data = 60;                        //再次赋值时,提示出错
    }
    public void function2(final Date date) {
        date = new Date();                //final 修饰引用形参,下面的赋值语句出错
        date.setYear(2020);               //final 修饰的引用参数的值是可以改变的
    }
}
```

3. final 方法

父类中的使用 final 修饰的方法在其子类中不能被重写,将方法定义为 final 类型可以防止任何子类修改该方法的定义与实现方式。

4. final 类

添加 final 修饰的类不能被继承,也就是不可以定义其子类,如 java.lang.String 及 java.lang.Math 类都是 final 类。如果将某个类设置为 final 形式,则类中的所有方法都被隐式设置为 final 形式,但是 final 类中的成员变量可以被定义为 final 或非 final 形式。

5.3.15 继承与组合

继承与组合是实现软件复用的两种重要手段。与继承不同,组合是使用已有类的对象作为实例变量生成新类,以达到软件复用的目的。例 5-6 示范了通过继承与组合实现代码的复用。

例 5-6

```
public class Person {
```

```java
    private String idNumber;              //身份证号码
    private String name;                  //姓名
    public String getIdNumber() {
        return idNumber;
    }
    public void setIdNumber(String idNumber) {
        this.idNumber = idNumber;
    }
    public String getName() {
        return name;
    }
    public void setName(String name) {
        this.name = name;
    }
}
public class Student extends Person{
    private String studentNo;             //学号
    private float score;                  //成绩
    public Teacher advisor;               //导师
}
public class Teacher extends Person{
    private String title;                 //职称
}
```

类 Student 及类 Teacher 继承自类 Person，子类 Student、Teacher 获得了父类 Person 的所有成员变量及方法（构造方法除外），这就是通过继承实现的代码复用。父类与子类之间体现的是一般与特殊的关系，强调"is-a"的关系。通过继承虽然很好地复用了父类的成员变量和方法，但同时也破坏了父类的封装性：子类可以直接访问父类的成员。

例 5-6 中，类 Student 中有一个指向 Teacher 对象的引用，这样一来，类 Student 与 Teacher 之间就具有了组合关系。通过 Teacher 对象引用 advisor，在 Student 类中可以复用 Teacher 类中定义的成员变量与方法。组合体现的是整体和部分，强调的是"has-a"的关系。在进行类的设计时，通常需要根据语义判断类之间是采用继承关系还是组合关系实现代码复用的。较继承而言，组合的实现方式更加简单，组合类与部分类之间相对独立，耦合松散。

5.4 个人通讯录（一）设计步骤

5.4.1 个人通讯录系统类图

根据个人通讯录（一）的功能要求，设计个人通讯录系统的类图，如图 5-3 所示。

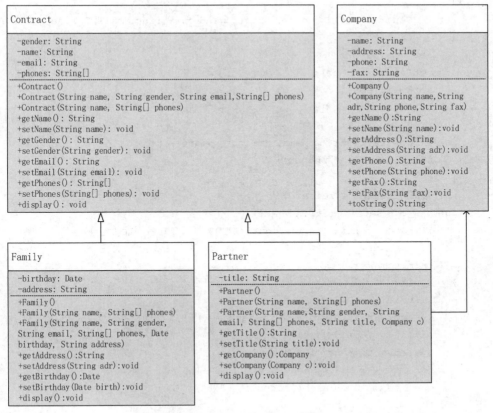

图 5-3 个人通讯录(一)类图

5.4.2 定义类 Contract

根据类图完成类 Contract 的定义。

(1) 包含姓名(name)、性别(gender)、电子邮件(email)、若干个联系电话(phones)等实例变量。

(2) 为每个实例变量生成 getter 方法及 setter 方法,其中姓名不能为空。

(3) 为联系人定义 3 个构造方法:默认构造方法、包含所有成员变量的构造方法、包含姓名及联系电话的构造方法。

(4) 定义 display()方法,输出成员变量的详细信息。

```
public class Contract {
    private String name;
    private String gender;
    private String email;
    private String[] phones;
    public Contract(){
```

```java
        }
        public Contract(String name, String gender, String email,String[] phones)
        {
            setName(name);
            setGender(gender);
            setEmail(email);
            setPhones(phones);
        }
        public Contract(String name, String[] phones) {
            this(name,"","",phones);
        }
        //省略部分 getter 及 setter 方法
        public void setName(String name) {
            if(name == null || name.equals(""))      //姓名不能为空
                return;
            this.name = name;
        }
        public void display() {
            System.out.println("姓名:"+getName()+"\t 性别:"+getGender()+"\te-
            mail:"+getEmail());
            System.out.print("联系电话: \t");
            for (int i = 0; i < phones.length; i++) {
                System.out.print(phones[i] +"\t");
            }
            System.out.println();
        }
    }
```

5.4.3 定义类 Family

5.4.3

根据继承关系定义家庭联系人类 Family。

(1) 除了从父类 Contract 继承的实例变量之外,类 Family 还需要定义家庭地址(address)及生日(birthday)这两个成员变量。

(2) 为每个成员变量生成 getter 及 setter 方法。

(3) 为类 Family 定义两个构造方法,通过调用父类的构造方法实现对象的初始化。

(4) 重写父类的 display()方法,实现子类对象的输出。

```java
public class Family extends Contract {
    private Date birthday;
    private String address;
    public Family() {
    }
    public Family(String name, String[] phones) {
```

```java
        super(name, phones);
    }
    public Family(String name, String gender, String email,
            String[] phones, Date birthday, String address) {
        super(name, gender, email, phones);
        setBirthday(birthday);
        setAddress(address);
    }
    public void display() {
        super.display();
        SimpleDateFormat sdf = new SimpleDateFormat("yyyy-MM-dd");
        System.out.println ("出生日期:"+sdf.format(getBirthday())
                        + "\t 家庭住址:"+getAddress());
    }
}
```

5.4.4 定义类 Partner

5.4.4

定义工作伙伴类 Partner 及其所在的公司类 Company。
(1) 根据图 5-3 实现公司类 Company 的定义。
(2) 通过继承定义合作伙伴类 Partner,添加其所在公司(company)及职称/职务(title)成员变量。
(3) 为每个成员变量生成 getter 及 setter 方法。
(4) 为类 Partner 定义两个构造方法,通过调用父类的构造方法实现对象的初始化。
(5) 重写父类的 display()方法,实现子类对象的输出。

```java
public class Company {
    private String name;
    private String address;
    private String phone;
    private String fax;
    public Company() {
    }
    public Company(String name, String address, String phone, String fax)
        setAddress(address);
        setPhone(phone);
        setFax(fax);
    }
    public String toString() {
        return String.format("% s\t 地址:% s\n 联系电话:% s\t 传真:% s\n",
                    getName(),getAddress(),getPhone(),getFax());
    }
}
```

```java
public class Partner extends Contract {
    private String title;
    private Company company;
    public Partner() {
    }
    public Partner(String name, String[] phones) {
        super(name, phones);
    }
    public Partner(String name, String gender, String email,
            String[] phones, String title, Company company) {
        super(name, gender, email, phones);
        setCompany(company);
        setTitle(title);
    }
    //省略 getter 与 setter 方法
    public void display() {
        super.display();
        System.out.println("职务:"+getTitle()+"\n 工作单位:"+getCompany());
    }
}
```

5.5 练一练

1. 定义课程信息类，包含课程编号、课程名称及学生成绩。编程实现对软件工程专业某班级的所有课程成绩进行统计，包括平均成绩、最高成绩、最低成绩，并打印成绩等级的分布律。

2. 运行下列代码，分析对象的初始化过程。

```java
public class InitializationClassTest {
    public static void main(String[] args) {
        SubClass sub1 = new SubClass();
        SubClass sub2 = new SubClass();
    }
}
class SuperClass{
    static {
        System.out.println("super class static initialization block.");
    }
    {
        System.out.println("super class instance initialization block.");
    }
    public SuperClass() {
        System.out.println("super class constructor method.");
```

```java
        }
    }
    public class SubClass extends SuperClass{
        static {
            System.out.println("sub class static initialization block.");
        }
        {
            System.out.println("sub class instance initialization block.");
        }
        public SubClass() {
            System.out.println("sub class constructor method.");
        }
    }
```

第6章 多态与内部类——个人通讯录（二）

6.1 个人通讯录（二）

设计个人通讯录（二）：通讯录（PhoneBook）用来保存若干联系人的信息且可以按照联系人姓名的拼音升序排列。要求通讯录提供以下功能：
(1) 添加联系人信息；
(2) 删除指定联系人信息；
(3) 查找指定联系人信息；
(4) 修改指定联系人信息；
(5) 显示所有联系人信息；
(6) 清空所有联系人。

6.2 程序设计思路

个人通讯录（二）综合应用了面向对象的三大特性：封装、继承及多态。个人通讯录（一）中已经实现了 Contract、Family、Partner 及 Company 类的定义，在对上述类进行重构的基础上，个人通讯录（二）增加了通讯录 PhoneBook 类，通过封装技术以联系人数组的形式保存若干联系人信息。联系人数组可以存储 Contract 类对象或者其子类 Family、Partner 对象。PhoneBook 类中还定义了若干方法，用来实现通讯录中联系人的增、删、改、查操作。通过多态技术，在方法中通过父类对象的引用，既可以动态绑定父类方法，也可以绑定子类中重写的父类方法，使应用程序具有良好的扩展性，可以在不改变原有代码的情况下扩展程序的功能。

6.3 关键技术

6.3.1 多态

1. 多态的概念

首先思考一个例子。主人（Master）养了猫（Cat）和狗（Dog）等多种宠物（Pet），并且每种宠物都有自己喜欢的食物（Food），如猫喜欢吃猫粮（CatFood）、狗喜欢吃骨头（Bone），那么如何实现主人给猫、狗等宠物喂食的行为呢？我们可以设想一下，在主人 Master 类中需要定义：

```
class Master{
```

```
public void feed(Cat cat, CatFood catFood){
    cat.eat(catFood);
}
public void feed(Dog dog, Bone bone){
    dog.eat(bone);
}
```

如果主人家又引入了新的宠物,那么就需要修改主人 Master 类,增加关于新的宠物的喂养方法,这样的实现思路显然不利于程序的扩展。考虑在 Master 类中定义一个"通用方法":

```
public void feed(Pet pet, Food food){
    pet.eat(food);
}
```

这样,即使再引入新的宠物,也无须修改 Master 类的定义。那么,如何实现这种"通用情况下的通用方法"呢?这就需要引入"多态"的概念。

所谓的多态,是指具有不同功能的方法可以使用相同的方法名,可以向不同的对象发送同一条消息(相同的方法名),不同的对象在接收时会产生不同的行为(即不同的方法实现)。利用多态性可以大大提高代码的扩展性及可维护性。

2. 引用类型的向上转型

要理解多态性,首先要知道什么是"向上转型"。向上转型是指父类引用变量指向子类对象,语法规范定义为:

父类类型 引用变量 = 子类类型对象;

转型之后的父类引用将不能访问子类特有的方法和数据,只能访问父类所具有的方法及变量,但是如果子类重写了父类方法,则运行时将按照重写的方法定义执行。

这种不同子类类型引用、调用同一个方法并呈现不同的行为特征就是多态。

例 6-1 示范了通过多态实现对"通用情况进行编程"的思想。

例 6-1

食物类及其子类的定义如下:

```
public class Food {                     //食物类
    private float weight;               //重量
    private String name;                //名称
}
public class CatFood extends Food{      //猫粮
    public String getName() {
        return "猫粮";
    }
}
```

```java
public class Bone extends Food{           //骨头
    public String getName() {
        return "骨头";
    }
}
```

宠物类及其子类的定义如下:

```java
public class Pet{                         //宠物类
    private int age;                      //年龄
    private String name;                  //名字
    public Pet(int age, String name) {
        this.age = age;
        this.name = name;
    }
    public void eat(Food food) {          //吃行为在宠物类中提供空实现
    }
}
public class Cat extends Pet{             //猫
    public Cat(int age, String name) {
        super(age, name);
    }
    public void eat(Food food) {
        if(food.getName().equals("猫粮"))
            System.out.println(this.getName()+"喜欢吃"+food.getName());
        else
            System.out.println(this.getName()+"不喜欢吃"+food.getName());
    }
}
public class Dog extends Pet{             //狗
    public Dog(int age, String name) {
        super(age, name);
    }
    public void eat(Food food) {
        if(food.getName().equals("骨头"))
            System.out.println(this.getName()+"喜欢吃"+food.getName());
        else
            System.out.println(this.getName()+"不喜欢吃"+food.getName());
    }
}
```

主人类的定义如下:

```java
public class Master{
    public void feed(Pet pet, Food food) {
        pet.eat(food);
```

```
    }
    public static void main(String args[]) {
        Master m = new Master();
        Pet cat = new Cat(1,"咪咪");            //上转型
        Pet dog = new Dog(2,"旺旺");
        m.feed(cat, new CatFood());
        m.feed(dog, new Bone());
    }
}
```

该程序的运行结果如下所示。

咪咪喜欢吃猫粮
旺旺喜欢吃骨头

若主人家新引入了宠物仓鼠(Hamster),仓鼠喜欢吃小米、大米等谷物(Grain),这时仅需要分别定义宠物子类(仓鼠类)及食物子类(谷物类)即可。

```
public class Grain extends Food{
    public String getName() {
        return "谷物";
    }
}
public class Hamster extends Pet{
    public Hamster(int age, String name) {
        super(age, name);
    }
    public void eat(Food food) {
        if(food.getName().equals("谷物"))
            System.out.println(this.getName()+"喜欢吃"+food.getName());
        else
            System.out.println(this.getName()+"不喜欢吃"+food.getName());
    }
}
```

3. 引用类型的向下转型

向下转型是指将父类的引用变量强制类型转换为子类类型,其语法规范为

子类类型 引用变量 = (子类类型)父类类型引用变量;

向下转型的前提:父类类型的引用变量必须是指向子类对象的,否则会抛出 ClassCaseException 异常信息。经过向下转型之后,就可以访问子类中新增加的成员了。

6.3.2 instanceof 运算符

instanceof 运算符用来判断对象是否是特定类或者子类的实例,返回 boolean 类型

值,其定义为

引用变量 instanceof 类型

instanceof 运算符通常用于向下转型之前,首先使用 instanceof 运算符判断一个对象是否是指定类型的实例,当运算结果为 true 时,才可以使用向下转型进行强制类型转换,否则会抛出 ClassCaseException 异常信息。例如,重新定义例 6-1 中的类 Cat:

```
public class Cat extends Pet{
    public Cat(int age, String name) {
        super(age, name);
    }
    public void eat(Food food) {
        if(food instanceof CatFood)    //判断 food 对象是否是 CatFood 类或者其子类实例
            System.out.println(this.getName()+"喜欢吃猫粮");
        else
            System.out.println("不是猫粮"+this.getName()+"不喜欢吃");
    }
}
```

6.3.3 抽象类与抽象方法

面向对象的概念中定义了两种类,分别为具体类和抽象类。前面我们定义的类都是具体类,那么什么是抽象类呢? 如果一个类只定义了一个为所有子类共享的一般形式,至于实现的细节则交给每一个子类完成,这种类不对应具体的实例,只具有一些抽象的概念,那么这样的类称为抽象类。

例如例 6-1 中的定义类 Pet,其中定义了方法 eat(Food food),说明每个宠物都具备"吃"的行为,具体如何实现这个行为需要在子类中给出具体定义,也就是类 Pet 无法准确地定义宠物的"吃"的行为。上述程序 Pet 类中对 eat(Food food)方法提供了空实现,如何避免定义空实现的方法呢? 这就需要引入抽象方法及抽象类。

抽象类及抽象方法的定义如下:

```
[public] abstract class 类名{
    ...
    //定义抽象方法
    [访问权限修饰符] abstract  返回值类型 抽象方法名(参数列表);
}
```

注意:抽象方法不能有方法体,即方法名后面没有大括号。

例如定义例 6-1 中的类 Pet:

```
public abstract class Pet{
    //省略其他方法定义
    public abstract void eat(Food food);
}
```

抽象类除了不能实例化对象之外，类的其他功能依然存在：成员变量、成员方法的访问方式和具体类相同。抽象类也可以定义初始化块及构造方法，用于在初始化子类时被调用。抽象类是对具体类的抽象，用来完成类框架共享的公共设计，具体子类继承并扩展公共设计。

6.3.4 接口

在 Java 语言规范中，接口是一系列抽象方法的声明，也就是接口中只有方法的特征而没有方法的实现，因此这些方法可以在不同的地方被不同的类实现，而这些实现可以具有不同的行为。接口可以理解为一种特殊的类，由全局常量和公共的抽象方法组成。接口是解决 Java 无法使用多继承的一种手段，但是接口在实际应用中更多的作用是制定一种协议，或者也可以把接口理解为不包含具体方法的抽象类。

1. 接口的定义

```
[public] interface 接口名 extends 父接口 1,父接口 2,…{
    常量的定义；
    抽象方法的定义；
}
```

注意：

(1) 接口与类不同，它可以有多个父接口；
(2) 接口中定义的方法默认具有 public、abstract 修饰符；
(3) 接口中定义的常量默认具有 public、static 及 final 修饰符。

接口与抽象类类似，是对类的一种抽象，指明了类所具备的功能，接口描述的是一种能力。例如组织一次会议有接待参会人员的需求，如招待不同身份的人的食宿问题，其对应规则如表 6-1 所示。

表 6-1 会议接待规则

参会人员	就餐地点	住宿地点
专家	酒店	酒店
教师	教工餐厅	招待所
学生	学生食堂	学生宿舍

例 6-2

定义接口 Receptional，提供为参会人员解决食宿问题的能力。

```
public interface Receptional{
    void eat();         //就餐
    void sleep();       //住宿
}
```

2. 子类实现接口

[修饰符] class 类名 extends 父类 implements 接口1,接口2,…{
    类的定义
}

子类中必须重写父接口中定义的所有抽象方法,否则该类只能定义为抽象类。

上述代码定义了接口 Receptional,下面就可以通过接口定义可接待的参会人员了。

```
class Expert implements Receptional{
    public void eat() {
        System.out.println("酒店就餐");
    }
    public void sleep() {
        System.out.println("酒店住宿");
    }
}
class Teacher implements Receptional{
    public void eat() {
        System.out.println("教工餐厅就餐");
    }
    public void sleep() {
        System.out.println("招待所住宿");
    }
}
class Student implements Receptional{
    public void eat() {
        System.out.println("学生食堂就餐");
    }
    public void sleep() {
        System.out.println("学生宿舍住宿");
    }
}
public class MeetingReception {
    public static void main(String[] args) {
        Receptional[] attendees = new Receptional[] {new Expert(),
                                    new Teacher(),new Student()};
        for(int i = 0; i < attendees.length; i++) {
            System.out.println("第"+(i+1)+"位参会人员:");
            attendees[i].eat();
            attendees[i].sleep();
        }
    }
}
```

3. 接口中的默认方法

在上述会议接待程序中，要求在已经做好的系统中添加展示参会人员身份证明的功能。如果直接在接口中添加新方法，那么需要对所有的实现类都重写新方法，这样做必然会影响已有的实现类，导致应用崩溃。为了解决这个问题，Java 8 在接口中引入了默认方法的定义。

默认方法的语法规范如下：

```
default 返回值类型 方法名(参数列表){
    方法体
}
```

默认方法总具有 public 访问权限，它由实现该接口的类实例调用。实现类可以继承接口中的默认方法，也可以重写接口中的默认方法。

下面就在接口 Receptional 中增加展示参会人员身份证明的功能。

```
public interface Receptional{
    ...
    default void showIdentityCard() {
        System.out.println("中国居民身份证");
    }
}
```

在接口 Receptional 增加默认方法 showIdentityCard()，并提供方法的实现，这样接口的所有实现类都自动继承了接口中的默认方法。在系统中增加外宾参会人员，定义如下：

```
class ForeignGuest implements Receptional{
    public void eat() {
        System.out.println("对外酒店就餐");
    }
    public void sleep() {
        System.out.println("对外酒店住宿");
    }
    public void showIdentityCard() {
        System.out.println("护照");
    }
}
```

对外宾人员重写 eat()、sleep() 及 showIdentityCard() 方法。

4. 接口中的静态方法

Java 8 在接口中引入了静态方法的定义，其语法规范如下：

```
static 返回值类型 方法名(参数列表){
```

方法体

}

接口中的静态方法具有 public 访问权限。静态方法由接口调用,格式为:"接口名.静态方法",无法通过其实现类访问。

在接口 Receptional 中增加静态方法,实现所有与会人员都具有相同的"统一接待程序",完整的接口 Receptional 定义如下:

```
public interface Receptional{
    void eat();           //就餐
    void sleep();         //住宿
    default void showIdentityCard() {
        System.out.println("中国居民身份证");
    }
    static void NormalRecept() {
        System.out.println("统一的接待程序:");
        System.out.println("发放会议证件及资料");
    }
}
```

定义测试程序如下:

```
public class MeetingReception {
    public static void main(String[] args) {
        Receptional[] attendees = new Receptional[] {new Expert(),
                new Teacher(),new Student(),new ForeignGuest()};
        for(int i = 0; i < attendees.length; i++) {
            System.out.println("第"+(i+1)+"位参会人员:");
            Receptional.NormalRecept();
            attendees[i].eat();
            attendees[i].sleep();
            attendees[i].showIdentityCard();
        }
    }
}
```

5. 接口与抽象类

接口与抽象类是两个很相似的概念,很容易被混淆,并且从代码实现的角度来看,很多时候,接口与抽象类也是可以互相替代的。但是,从设计目的及使用动机而言,两者存在很大的差别。

在抽象类的类体中,除了可以包含抽象方法之外,它与普通类是相同的,可以包含构造方法、初始化块、类变量、类方法、实例变量及实例方法;而接口中仅能包含静态常量、抽象方法的定义。Java 8 之后,接口中又增加了默认方法及静态方法的定义。

从与类的关系来看,一个类只能继承一个抽象类,而一个类却可以实现多个接口。

除了以上差别之外，抽象类更主要的是体现了类模板的设计思想，定义其子类的通用特性及部分已经实现的共有功能。抽象类与子类之间体现了"一般与特殊"之间的关系。而接口则是规则的集合，是实现类必须遵守的设计规范。从应用程序架构的角度来看，接口体现了多模块之间的耦合标准。

6.3.5 内部类

定义在其他类内部的类称内部类，包含内部类的类也称为外部类。那么，为什么需要定义内部类呢？我们知道，类是对同种事物的一种抽象，当我们描述事物时，如果事物中还包含另一种事物，而且被包含的事物还需要了解其外部包含事物的特性，并能与之通信，这时就需要定义内部类以描述及抽象被包含事物。例如，汽车通常由发动机、底盘、车身和电气设备这4大部分组成，汽车的发动机只能在汽车中使用，并且发动机需要依赖汽车的其他组件才能正常运转，所以我们应该定义发动机为汽车的内部类。

相较外部类，内部类具有如下特点：

(1) 内部类中可以直接访问外部类的成员，包括私有成员，但是外部类不能直接访问内部类的成员；

(2) 内部类虽然在类的内部定义，但它是一个独立的类。经过编译，内部类会被编译成独立的字节码文件，字节码文件命名格式为：外部类名$内部类名.class。如在外部类Car中定义内部类Engine，经过编译之后，会生成Car.class、Car$Engine.class两个字节码文件；

(3) 内部类除了可以使用public与缺省的访问修饰符之外，还可以使用private、protected及static等访问修饰符。

例6-3示范了内部类的定义。

例 6-3

```
public class Car {
    private boolean isFailure;              //是否故障
    private Engine engine = new Engine(3.0d); //发动机
    public void setFailure(boolean isFailure) {
        this.isFailure = isFailure;
    }
    private class Engine{                   //定义内部类
        private double capacity;            //发动机的排量
        public Engine(double capacity) {
            this.capacity = capacity;
        }
        public void fireUp() {
            if(!isFailure)                                    ①
                System.out.println("发动机点火");
            else
                System.out.println("汽车故障,发动机无法点火");
```

```
        }
    }
    public void start() {
        System.out.println("启动汽车");
        engine.fireUp();
    }
    public void display() {
        System.out.println("汽车状态:"+(isFailure?"故障":"正常"));
        //System.out.println("发动机排量:"+capacity);           ②
        System.out.println("发动机排量:"+engine.capacity);      ③
    }
}
```

例 6-3 在外部类 Car 中定义了类 Engine,Engine 类与外部类的成员变量和成员方法类似,所以它是 Car 类的内部类。程序经过编译,在项目所在的 bin 目录下生成了两个字节码文件:Cat.class、Cat $ Engine.class。

内部类 Engine 中定义了成员变量 capacity 与成员方法 fireUp(),成员方法 fireUp() 代码行①处就是内部类中直接访问外部类的私有成员变量 isFailure。虽然内部类可以无条件地访问外部类的成员(包括私有成员),但是外部类却不能直接访问内部类的成员。如程序中代码行②处,如果去掉程序注释,则编译出错,提示"capacity 不能被解析为变量"。在外部类中,如果要访问内部类的成员,则可以通过内部类的对象进行访问,如程序中代码行③。engine 对象是外部类中 Engine 类型的成员变量。

6.3.6 匿名内部类

如果一个内部类在整个操作中只使用一次,则定义其为匿名内部类。顾名思义,匿名内部类就是没有名字的内部类。定义匿名内部类的语法规范如下:

```
new 父类构造方法(参数列表)|接口{
//匿名内部类的类体
}
```

例 6-3 中定义了如下接口:

```
public interface Audio{
    void turnOn();
    void trunOff();
}
```

在 Car 类中定义方法 audioTest()如下:

```
public class Car {
    public void audioTest() {
        String name = "春雷音响";                              ①
        Audio audio= new Audio() {                             ②
```

```
            private int volum;        //匿名类中定义实例变量              ③
            {        //匿名类中定义初始化块,实现实例变量的初始化          ④
                volum = 10;
            }
            @Override
            public void turnOn() {                                        ⑤
                //对外部类中的局部变量赋值,引发编译错误
                name = "匿名";                                             ⑥
                System.out.println(name+"已经打开,当前音量为:"+volum);    ⑦
            }
            @Override
            public void trunOff() {                                       ⑧
                volum= 0;
                System.out.println(name+"已经关闭,当前音量为:"+volum);
            }
        };
        audio.turnOn();          //外部类中使用匿名类实例对象
        audio.trunOff();
    }
}
```

匿名内部类继承自一个父类或者实现一个接口,定义中无 class 关键字。如本例中代码行②定义了一个实现接口 Audio 的匿名类,上述定义实际包括两部分:匿名类的定义及实例化匿名类对象。

匿名内部类的类体中必须重写接口或者其抽象父类中定义的所有抽象方法,也可以重写父类中定义的普通方法,如代码行⑤、⑧;匿名类中还可以定义实例变量,如代码行③,但是不能定义静态变量和静态方法。

匿名内部类的类体中无法定义构造方法,但是可以通过初始化块的方式实现初始化操作,如上例中的代码行④。

匿名内部类如果需要访问外部类中定义的局部变量,则该局部变量默认定义为 final 类型。Java 8 之前需要显式标注 final 类型。如上例中的代码行①定义了局部变量 name 并赋值,代码行⑦在匿名类中可以访问局部变量,但是无法对 final 类型的局部变量再次赋值,如代码行⑥会引发编译错误。

6.4 个人通讯录(二)设计步骤

6.4.1 系统类图

根据个人通讯录(二)的功能要求设计个人通讯录系统的类图,联系人 Contract 类与通讯录 PhoneBook 类之间构成聚合关系,如图 6-1 所示。

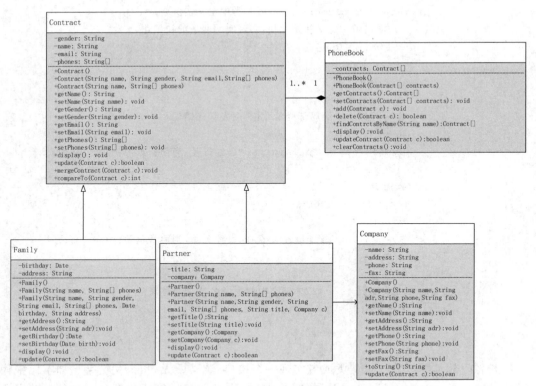

图 6-1 个人通讯录(二)类图

6.4.2 重构类 Contract

根据任务要求，通讯录按照联系人姓名的拼音首字母升序排列，重新定义联系人 Contract 类。

1. 添加方法实现按照联系人姓名进行比较

Collator 类用来执行区分语言环境的字符串比较。通过 Collator 类提供的静态工厂方法 getInstance()即可获得"中国"语言环境对应的 Collator 对象。

```
public int compareTo(Contract o) {
    Collator instance = Collator.getInstance(java.util.Locale.CHINA);
    return instance.compare(this.getName(),o.getName());
}
```

2. 合并同一联系人的不同信息

对同一联系人的电话号码进行合并时需要进行去重处理。

```
public void mergeContract(Contract o) {
```

```java
if(this.getName().equals(o.getName())) {
    if(this.getGender().equals("") )
        this.setGender(o.getGender());
    if(this.getEmail().equals(""))
        this.setEmail(o.getEmail());
    //复制并去重
    boolean flag;
    String[] newPhones = new String[o.getPhones().length];
    int count = 0;
    for(intj = 0; j < o.getPhones().length; j++) {
        flag = true;
        for(int i = 0; i < this.getPhones().length; i++) {
            if(o.getPhones()[j].equals(this.getPhones()[i])) {
                flag = false;
                break;
            }
        }
        if(flag)
            newPhones[count++] = o.getPhones()[j];
    }
    int position = phones.length;
    phones = Arrays.copyOf(phones, phones.length+count);         //数组扩容
    System.arraycopy(newPhones, 0, phones, position, count);     //追加元素
}
}
```

3. 定义方法实现联系人信息的修改

```java
public boolean update(Contract c) {          //修改当前联系人信息
    if (getName().equals(c.getName())) {
        if (c.getEmail() !=null && c.getEmail() !="")
            setEmail(c.getEmail());
        if (c.getGender() !=null && c.getGender() !="")
            setGender(c.getGender());
        if (c.getPhones() !=null && c.getPhones().length !=0)
            setPhones(c.getPhones());
        return true;
    } else
        return false;
}
```

6.4.3 重构类 Family

重载父类的 update(Contract c)方法,实现 Family 实例对象的修改。

6.4.3

```java
public boolean update(Contract c) {
    if(getName().equals(c.getName())) {
        if(c instanceof Family) {
            super.update(c);
            Family f = (Family)c;
            if(f.getBirthday()!=null)
                setBirthday(f.getBirthday());
            if(f.getAddress()!=null && f.getAddress()!="")
                setAddress(f.getAddress());
            return true;
        }else if(c instanceof Partner)
            return false;
        else {
            super.update(c);
            return true;
        }
    }else
        return false;
}
```

6.4.4 重构类 Partner

6.4.4

重载父类的 update(Contract c)方法，实现 Partner 实例对象的修改。

```java
public boolean update(Contract c) {
    if(getName().equals(c.getName())) {
        if(c instanceof Family) {
            return false;
        }else if(c instanceof Partner) {
            super.update(c);
            Partner p = (Partner)c;
            if(p.getTitle()!=null && p.getTitle()!= "")
                setTitle(p.getTitle());
            if(p.getCompany()!=null)
                getCompany().update(p.getCompany());
            return true;
        }
        else {
            super.update(c);
            return true;
        }
    }else
        return false;
}
```

6.4.5 重构类 Company

定义方法 update(Company c)方法，实现公司信息的修改。

```
public boolean update(Company c) {
    if(c == null)
        return false;
    if(c.getName()!=null && c.getName()!="")
        setName(c.getName());
    if(c.getAddress()!=null && c.getAddress()!="")
        setAddress(c.getAddress());
    if(c.getPhone()!=null && c.getPhone()!="")
        setPhone(c.getPhone());
    if(c.getFax()!=null && c.getFax()!="")
        setFax(c.getFax());
    return true;
}
```

6.4.6 定义通讯录 PhoneBook 类

通讯录 PhoneBook 中保存了若干联系人信息，需要提供通讯录的增、删、改、查等操作。

1. 封装联系人信息

PhoneBook 类中定义联系人 Contract 数组：

```
private Contract[] contracts;
```

保存联系人信息。此处联系人数组 contracts 可以存储 Contract 类对象或者其子类 Family、Partner 对象。

2. 定义 setter 方法

在 setContracts()方法中需要对联系人数组进行排序操作，以保证 PhoneBook 类按照联系人姓名的拼音升序排列。

```
public void setContracts(Contract[] contracts) {
    this.contracts = contracts;
    for (int i = 0; i < contracts.length - 1; i++) {      //对联系人进行排序
        Contract temp;
        for (int j = i + 1; j < contracts.length; j++)
            if (contracts[i].compareTo(contracts[j]) > 0) {
                temp = contracts[i];
```

```
            contracts[i] = contracts[j];
            contracts[j] = temp;
        }
    }
}
```

3. 添加联系人

区别 3 种情况分别处理。
- 空通讯录：分配数组空间，添加新的联系人。
- 添加通讯录中不存在的联系人：对数组进行扩容，添加新的联系人。
- 添加通讯中存在的联系人：对同一联系人进行合并操作。

```
private int findContract(Contract c) {
    for(int i = 0; i < contracts.length; i++) {
        if(contracts[i].getName().equals(c.getName())) {
            return i;
        }
    }
    return -1;
}
public void add(Contract c) {
    if(contracts = = null) {                          //空通讯录
        contracts = new Contract[1];
        contracts[0] = c;
        return;
    }
    int index = findContract(c);
    if(index < 0) {                                   //添加操作
        Contract[] contractAdded = Arrays.copyOf(contracts, contracts.length+1);
        contractAdded[contractAdded.length-1] = c;
        setContracts(contractAdded);
        return;
    }else
        contracts[index].mergeContract(c);            //合并操作
}
```

4. 删除联系人

查询联系人，通过移动数组元素实现元素的删除。

```
public boolean delete(Contract c) {
    int index = findContract(c);
    if(index < 0)
        return false;
    Contract[] contractDeleted = new Contract[contracts.length-1];
    System.arraycopy(contracts, 0, contractDeleted, 0, index);
```

```
        System.arraycopy(contracts, index+1, contractDeleted, index,contracts.
                    length-1-index);
        contracts = contractDeleted;
        return true;
}
```

5. 根据姓名进行模糊查询

利用字符串的 contains() 方法对字符串进行模糊查询,获取满足条件的所有联系人。

```
public Contract[] findContractsByName(String name) {
    Contract[] result = new Contract[contracts.length];
    int num = 0;
    for(int i = 0; i < contracts.length; i++) {
        if(contracts[i].getName().contains(name))
            result[num++] = contracts[i];
    }
    return Arrays.copyOf(result, num);
}
```

6. 修改联系人信息

首先查询待修改的联系人,然后利用多态技术修改联系人信息。

```
public boolean updateContract(Contract c) {
    int index = findContract(c);
    if (index < 0)
        return false;
    contracts[index].update(c);
    return true;
}
```

7. 显示所有联系人信息

遍历联系人数组,利用多态技术调用 display() 方法,输出联系人或者子类对象的信息。

```
public void display() {
    for(int i = 0; i < contracts.length; i++) {
        contracts[i].display();
    }
}
```

8. 清空所有联系人

```
public void clearContracts() {
    contracts = null;
}
```

6.5 练一练

1. 分析下列程序的运行结果。

```java
public interface USB {
    default void work() {
        System.out.println(this.getName()+"在工作");
    };
    String getName();
    default void popUp() {
        System.out.println(this.getName()+"已拔出");
    };
}
public class Phone implements USB {
    public String getName() {
        return "Phone.";
    }
}
public class Camera implements USB {
    public String getName() {
        return "Camera.";
    }
}
public class USBKey implements USB {
    public String getName() {
        return "USB key";
    }
}
public class Computer {
    private String cpu;
    private String mainCard;
    private USB usb;
    public void commByUSB() {
        usb.work();
    }
    public void popUpUSB() {
        usb.popUp();
    }
    //省略 getter、setter 及构造方法
    public static void main(String[] args) {
        Computer c = new Computer("Intel","ASUS");
        USB phone = new Phone();
        USB useKey = new USBKey();
```

```
        USB camera = new Camera();
        c.setUsb(phone);
        c.commByUSB();
        c.popUpUSB();
        c.setUsb(useKey);
        c.commByUSB();
        c.popUpUSB();
        c.setUsb(camera);
        c.commByUSB();
        c.popUpUSB();
    }
}
```

2. 阅读下列程序,分析程序的运行结果。

(1) 抽象类 MobilePhone。

```
public abstract class MobilePhone {
    private String brand;
    //省略 getter setter 及构造方法
    public abstract void call(String phoneNumber);
    public abstract void sendMessage(String phoneNumber,String mess);
}
```

(2) 抽象类 Camera。

```
public abstract class Camera {
    public abstract void takeAPicture();
}
```

(3) 接口 FliterLens。

```
public interface FilterLens {          //滤镜接口
    void filter();
}
```

(4) 类 SmartPhone。

```
public class SmartPhone extends MobilePhone{
    class CameraPro extends Camera implements FilterLens{
        @Override
        public void filter() {
            System.out.println("启动滤镜功能");
        }
        @Override
        public void takeAPicture() {
            System.out.println("照相");
        }
```

```java
    }
    private CameraPro camera;
    public SmartPhone(String brand) {
        this.setBrand(brand);
        camera = new CameraPro();
    }
    public void takeAPicture() {
        camera.filter();
        camera.takeAPicture();
    }
    @Override
    public void call(String phoneNumber) {
        System.out.println("给"+phoneNumber+"打电话");
    }
    @Override
    public void sendMessage(String phoneNumber,String mess) {
        System.out.println("给"+phoneNumber+"发消息"+mess);
    }
    public static void main(String[] args) {
        SmartPhone phone = new SmartPhone("华为 P40");
        phone.call("15066385530");
        phone.sendMessage("15066385530", "近来好吗?");
        phone.takeAPicture();
    }
}
```

第7章 Java 常用类与枚举类——21点游戏

7.1 21点游戏介绍

21点是一个古老的纸牌游戏,起源于法国,后传入英国并广泛流传。游戏的英文名字为 BlackJack。游戏的基本规则是:玩家设法使手中所有纸牌的点数之和达到或尽可能接近21点,但是不能超过21点。

1. 游戏的具体规则

(1) 纸牌。

纸牌数:一副扑克牌,去除大小王,共52张纸牌。

花色:红桃、黑桃、方块、梅花。

纸牌的点数:A 到10的纸牌分别对应点数1到10,纸牌 J、Q、K 的点数皆为10。

(2) 游戏流程。

游戏开始,首先洗牌,然后依次为多名玩家发牌,玩家计算自己手中纸牌的点数和,如果点数和大于21点,则"爆牌(Bust)",直接判定该玩家失败,被淘汰出局。如果没有爆牌,则可以选择继续游戏、停牌及结束游戏这三项操作。如果玩家选择停牌,则后续游戏不再给该玩家发牌。

2. 游戏胜负判定规则

如果玩家选择结束游戏,则游戏点数最接近21点的玩家获胜;如果多位玩家的点数和相等或者最后同时爆牌,则平局。

7.2 程序设计思路

21点游戏包含纸牌、玩家、游戏等实体,分析游戏规则,可以初步得到系统类图,如图 7-1 所示(省略类的属性及方法定义)。

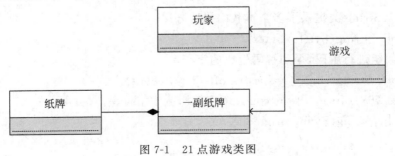

图 7-1 21点游戏类图

游戏类实现对玩家及一副纸牌的封装,提供游戏规则及胜负的判定。玩家类定义玩家姓名等属性,提供玩家的出牌策略。一副纸牌保存52张纸牌信息,提供洗牌、发牌等方法。所谓的洗牌,就是将一副牌的顺序打乱,其基本思想是通过迭代的方式从一副牌中随机取出一张牌与当前纸牌进行交换。这里可以引入 Random 类,用于随机抽取一张纸牌。另外,为了提高程序的可读性,纸牌的花色及牌面值可以定义为枚举类。

7.3 关键技术

7.3.1 Object 类

java.lang.Object 类是所有类的父类,即如果一个类在定义时没有显式声明父类,那么默认其父类就是 Object 类。所以,任何引用数据类型的对象(包括数组、类对象等)都可以经过向上转型赋值给 Object 类型的变量,并且也都可以调用 Object 类中声明的方法。

Object 类提供如下常用方法。

(1) public String toString()。

返回对象的字符串表示。通常子类中需要重写 toString()方法,以获得对当前对象的字符串描述信息。

(2) public int hashCode()。

返回该对象的哈希值。默认情况下,该方法会根据对象的物理地址计算哈希值。如果子类重写该方法,则要保证不同对象的 hashCode()值是不同的。子类中与 hashCode()方法同时重写的方法还有 equals()方法,用来确保相等的两个对象拥有相等的哈希值。

(3) public boolean equals(Object obj)。

判断 obj 对象与当前对象是否指向同一个引用。

7.3.2 String 类

字符串就是一个字符序列,在 Java 程序中,常使用字符串存储及处理文本。

1. 构造方法

java.lang.String 类提供了多个重载的构造方法。

(1) public String(byte[] bytes)。

对字节数组进行解码,得到解码生成的字符串。

(2) public String(char[] value, int offset, int count)。

将字符数组从 offset 开始的 count 个字符构造成字符串。

(3) public String(String original)。

```
public String(StringBuffer buffer)
```

```
public String(StringBuilder builder)
```

分别由 String、StringBuffer 及 StringBuilder 类对象构造字符串副本。例如：

```
String str1 = new String("original");
String str2 = new String(str1);                ①
System.out.println(str1==str2);      //输出 false
String str3 = "original";                      ②
String str4 = "original";                      ③
System.out.println(str3==str4);      //输出 true
System.out.println(str1==str3);      //输出 false
```

代码行①产生 str1 的字符串副本，引用赋值给字符串变量 str2，所以 str1 与 str2 拥有相同的字符串值，但是引用不同。

代码行②、③将字符串常量的引用赋值给字符串变量。关于字符串常量，Java 虚拟机为了提升性能和减少内存开销，避免字符串的重复创建，其内部维护了一块特殊的内存空间，即字符串池(String Pool)。当以代码行②、③方式创建一个字符串时，JVM 首先会去字符串池中查找是否存在"original"对象，如果不存在，则在字符串池中创建"original"对象，然后将池中 original 对象的引用地址返回给字符串 str3，这样 str3 就会指向池中"original"这个字符串对象；如果存在，则不创建任何对象，直接将池中"original"对象的地址返回并赋给字符串变量 str4，所以 str3 与 str4 拥有相同的引用。当采用 new 关键字创建 String 对象时，每次实例化都将产生一个新的对象，也就是说，引用 str3 和 str1 指向的是两个不同的对象。

2. String 类的常用方法

java.lang.String 类提供如下常用的字符串操作方法。

(1) public char charAt(int index)。

返回指定索引的字符。

(2) public int indexOf(int ch)。

返回字符 ch 在当前字符串中第一次出现的索引，如果当前字符串不包含字符 ch，则返回-1。

(3) public int compareTo(String anotherString)。

按照字典序比较 anotherString 与当前字符串，如果当前字符串位于 anotherString 字符串之前，则返回负整数；如果当前字符串位于 anotherString 字符串之后，则返回正整数；如果两个字符串的字符序列相等，则返回 0。

(4) public String concat(String str)。

连接当前字符串与字符串 str。与字符串运算符"+"的功能相似。

(5) public boolean equals(Object anObject)。

判断当前字符串与对象 anObject 是否包含相同的字符序列，相同则返回 true，否则返回 false。

（6）public boolean startsWith(String prefix)。

判断当前字符串是否具有前缀 prefix。

（7）public boolean endsWith(String suffix)。

判断当前字符串是否具有后缀 suffix。

（8）public boolean contains(CharSequence cs)。

判断当前字符串是否包含字符序列 cs。CharSequence 是一个描述字符串结构的接口，类 String、StringBuilder 及 StringBuffer 都实现了该接口。

（9）public int length()。

返回当前字符串的长度。

（10）public static String valueOf(XXX x)。

实现由基本数据类型构建字符串的操作。其中，XXX 代表基本数据类型，x 代表变量，如 int x 等。例如：

```
String s=String.valueOf(123);          //等价于 s="123"
```

（11）public String[] split(String regex)。

根据格式字符串 regex 拆分当前字符串，返回拆分后得到的字符串数组。

注意：若直接以"."""\""|"""""等作为分隔，则不能正确分隔源字符串，必须使用对应的转义字符"\\.""\\\\""\\|""\""。例如：

```
String IPAddress = new String("192.168.1.248");
String[] strs = IPAddress.split("\\.");
for(String ss : strs)
    System.out.println(ss);
```

（12）public String substring(int beginIndex, int endIndex)。

获取当前字符串从 beginIndex 开始到 endIndex－1 结束的子串。

7.3.3　StringBuilder 类与 StringBuffer 类

String 在 Java 中被定义为不变类，当 String 实例参与字符串的运算时，会产生很多中间字符串。Java 中的 StringBuffer 与 StringBuilder 类可以构建可变的字符序列，能够很好地解决这个问题。StringBuffer 和 StringBuilder 实例可以被多次修改，并且不产生新的中间字符串。

StringBuilder 类在 Java 5 中被首次提出，与类 StringBuffer 的操作基本相似。两者最大的不同在于 StringBuilder 的方法不是线程安全的（关于多线程的具体内容请参见第 13 章）。由于 StringBuilder 相较于 StringBuffer 操作速度较快，所以通常建议使用 StringBuilder 类。但是在多线程环境下，为了保证线程安全性，必须使用 StringBuffer 类。本章以 StringBuilder 类为例，讲解如何实现字符串的追加、插入、修改等操作。

1. 构造方法

java.lang.StringBuilder 类提供了多个重载方法，构建可变的字符序列。

(1) public StringBuilder(int capacity)。

构建长度为 0 的可变字符序列,初始容量为 capacity,省略参数,初始容量为 16。

注意:类 StringBuilder 有两个属性,分别为 length 与 capacity。属性 length 表示 StringBuilder 实例中实际包含的字符序列个数;属性 capacity 表示为 StringBuilder 实例可容纳的最大字符数。当 StringBuilder 实例存放的字符序列大于 capacity 所定义的字符数时,其容量会自动增加。

(2) public StringBuilder(String str)。

构建可变字符序列,字符序列初始化为字符串 str。

2. StringBuilder 类的常用方法

java.lang.StringBuilder 类提供了许多重载的插入、追加、删除和改变字符序列的方法。方法定义中,XXX 代表数据类型,x 代表变量,如 int x 等。

(1) public StringBuilder append(XXX x)。

将指定类型的数据追加到当前字符序列之后。

(2) public StringBuilder delete(int start, int end)。

删除可变字符序列从 start 开始到 end－1 结束的若干字符。

(3) public StringBuilder replace(int start, int end, String newStr)。

采用字符串 newStr 替换字符序列从 start 开始至 end－1 结束的字符序列。

(4) public void setCharAt(int index, char ch)。

采用字符 ch 替换当前字符序列中索引 index 处的字符。

(5) public String toString()。

返回当前可变字符序列的字符串表示。

7.3.4 Math 类

java.lang.Math 类提供了许多重载的静态数学运算方法。Math 类中还定义了两个静态常量 E 与 PI,其中,E 表示自然对数的底数 e;PI 表示圆周率。

Math 类的常用方法如下,方法定义中,XXX 代表数据类型,x 代表变量,如 int x 等。

(1) public static XXX abs(XXX x)。

返回参数 x 的绝对值。

(2) public static double ceil(double b)。

返回 double 类型数据,其值为大于或等于参数 b 的整数,也就是对参数 b 向上取整。

例如:

```
double d1 = 12.16;
double d2 = Math.ceil(d1);              //d2 = 13.0
```

(3) public static double floor(double b)。

返回 double 类型数据,其值为小于或等于参数 b 的整数,也就是对参数 b 取整。

(4) public static long round(double b)、public static int round(float b)。

返回最接近参数 b 的整数,即对参数 b 进行四舍五入操作。

(5) public static XXX max(XXX x1,XXX x2)。

返回两个指定类型数据中的最大值。

(6) public static XXX min(XXX x1,XXX x2)。

返回两个指定类型数据中的最小值。

(7) publicstatic double pow(double a,double b)。

计算 a^b。

(8) public static double random()。

返回一个位于区间[0.0,1.0]上且满足均匀分布的伪随机数。

7.3.5 Random 类

Java 提供了灵活的生成随机数的类 java.util.Random,可以方便地生成各种数据类型的随机数。

1. 构造方法

```
public Random()
public Random(long seeds)
```

Random 类提供了上述两个构造方法,用于构造伪随机数发生器。构造方法中可以通过参数 seeds 指定初始化的种子,也可以采用默认种子,即以当前系统时间作为随机数初始化的种子。

2. 随机数生成方法

通过 Random 实例对象可以生成各种数据类型的随机数。

(1) public double nextDouble()。

返回 double 类型、取值范围在[0.0,1.0)的一个伪随机数。

(2) public float nextFloat()。

返回 float 类型、取值范围在[0.0,1.0)的一个伪随机数。

(3) public int nextInt()。

获取一个伪随机整数,其取值范围是所有整数,即[-2^{31},$2^{31}-1$]。

(4) public int nextInt(int n)。

获取一个伪随机整数,其取值范围是[0,n−1]。

7.3.6 Date 类

Date 类是定义于 Java 1.0 版本的 java.util.Date 类,提供日期型数据的表达及操作。

1. 构造方法

(1) public Date()。

实例化日期对象，表达当前的系统日期及时间。

(2) public Date(long date)。

实例化日期对象，参数指代从 1970 年 1 月 1 日 00:00:00 GMT 开始经历的毫秒数。

2. Date 类的常用方法

(1) public long getTime()。

获取指定时间实例对应的 long 型整数（即从 1970 年 1 月 1 日 00:00:00 GMT 开始到指定时间实例经历的毫秒数）。

(2) public void setTime(long n)。

设置指定时间实例对应的 long 型整数。

7.3.7 Calendar 类

Java 1.1 之后推出了 java.util.Canlendar 类，该类可以方便地实现日历的表达及操作。

1. 构建 Canlendar 类实例

Canlendar 定义为抽象类，不能直接实例化对象，可以通过以下静态方法获取当前系统的日期实例。

```
public static Calendar getInstance()
```

2. set()方法与 get()方法

Calendar 类中定义了整型的静态变量，用于表示特定的日历字段，如 Calendar.YEAR、Calendar.MONTH 等。

通过如下 set()方法和 get()方法可以设置及获取日历字段的值。

```
int get(int field)
void set(int field, int value)
```

例如，假设当前系统日期是 2020 年 5 月 26 日。

```
Calendar c = Calendar.getInstance();
int year = c.get(Calendar.YEAR);        //year= 2020
int month = c.get(Calendar.MONTH);      //month= 4
int day = c.get(Calendar.DATE);         //day= 26
```

3. Calendar 类实例与 Date 类实例之间的转换

Calendar 类与 Date 类实例之间可以通过如下方式相互转换。例如：

```
Date d = new Date();
Calendar cc = Calendar.getInstance();
cc.setTime(d);        //Date 类实例转换为对应日期的 Calendar 类实例
d.setTime(cc.getTimeInMillis());    //Calendar 类实例转换为对应日期的 Date 类实例
```

7.3.8 SimpleDateFormat 类

java.text.SimpleDateFormat 类可以实现 Date 实例的格式化,也可以解析各种样式的日期字符串,即实现 Date 实例与 String 类型数据的相互转换。

1. Date 实例的格式化

Date 实例的格式化问题也就是将 Date 类型数据转换为 String 类型数据。首先需要构造 Java.text.SimpleDateFormat 类的实例以实现格式化。

```
public SimpleDateFormat(String pattern)
```

其中,参数 pattern 是自定义的格式字符串。常用的格式字符串"yyyy-MM-dd HH:mm:ss"表示"年-月-日 时:分:秒",例如"2020-05-25 12:03:05"。关于格式字符的详细说明请参见 Java API 文档。

SimpleDateFormat 类提供了 format 方法,根据 SimpleDateFormat 类实例中定义的格式将一个 Date 类型转换成 String 类型数据:

```
public String format(Date date)
```

2. String 实例的解析

String 实例的解析问题也就是将 String 类型数据转换为 Date 类型数据。SimpleDateFormat 类提供了 parse 方法,根据 SimpleDateFormat 类实例中定义的格式将一个 String 类型数据转换为 Date 类型数据:

```
public Date parse(String text)
```

7.3.9 枚举类

在某些情况下,一个类的对象是有限且固定的。例如描述纸牌的花色,它只有 4 个对象,这种对象实例拥有有限且固定的类,在 Java 中可以定义枚举类型进行描述。以下代码是枚举类的一个简单应用。

```
public enum SuitEnum {
    SPADE,HEART,DIAMOND,CLUB;
}
public class SuitEnumTest {
    public staticvoid main(String[] args) {
```

```
        SuitEnum suit = SuitEnum.HEART;
        System.out.println(suit.name());        //输出：HEART
    }
}
```

代码中使用关键字 enum 定义枚举类 SuitEnum。SuitEnum 的第一行定义了 4 个枚举类实例。在 SuitEnumTest 类中，定义 SuitEnum 类型的变量 suit 获取 SuitEnum.HEART 实例的引用。调用 suit.name()方法获得枚举对象的名称，即 HEART。

1. 枚举类的定义

枚举类型在 Java 中是一种引用类型，是一种特殊的类。枚举类型中定义所有的实例对象，并且实例对象的定义必须放在类定义的第一行，系统自动添加 public static final 修饰符。实例对象的构造无须使用 new 关键字显式调用构造方法。

在枚举类型中，除了定义枚举实例外，还可以定义成员变量和方法，也可以定义构造方法。例 7-1 对上述 SuitEnum 类的定义进行了重构，增加了成员变量、方法及构造方法的定义。

例 7-1

```java
public enum SuitEnum {
    SPADE("黑桃"),HEART("红桃"),DIAMOND("方块"),CLUB("梅花");
    private final String name;
    SuitEnum(String name){
        this.name = name;
    }
    public String getName() {
        return name;
    }
}
public class SuitEnumTest {
    public static void main(String[] args) {
        SuitEnum suit = SuitEnum.HEART;
        System.out.println(suit.name());            //输出：HEART
        System.out.println(suit.getName());         //输出：红桃
    }
}
```

例 7-1 枚举类 SuitEnum 中定义了 final 类型的成员变量 name，在构造方法中为成员变量 name 赋值，并且 SuitEnum 实例对象的 name 属性是不可更改的。

2. 枚举类型与类类型的区别

虽然枚举类型与类类型很相似，但两者也存在一些差别。
（1）枚举类默认继承了 java.lang.Enum，而不是继承 Object 类。
枚举类型已经具有父类 Enum，所以不能再显式继承其他父类，但是枚举类型可以实现一个或者多个接口。

（2）定义非抽象的枚举类默认使用 final 修饰。

非抽象的枚举类型不能作为父类派生子类类型。

（3）枚举类的构造器只能被定义为 private。

因为枚举类型是一种对象实例有限且固定的类，所以不需要也无法在枚举类的外部构建新的枚举实例。

7.4 21点游戏设计步骤

7.4.1 纸牌类

根据纸牌的特性设计 Card 类。每张纸牌具有花色（suit）与牌面值（face）两个属性，并且花色与牌面值属性具有有限且固定的实例对象，所以将其定义为枚举类。

1. 枚举类 SuitEnum

定义枚举类 SuitEnum，描述纸牌的 4 种花色：红桃、黑桃、方块及梅花。

```java
public enum SuitEnum {
    SPADE("黑桃"),HEART("红桃"),DIAMOND("方块"),CLUB("梅花");
    private final String name;
    SuitEnum(String name){
        this.name = name;
    }
    public String getName() {
        return name;
    }
}
```

2. 枚举类 FaceEnum

定义枚举类 FaceEnum 描述纸牌 A～K 共 13 种牌面值。FaceEnum 中定义 name 及 point 属性，分别说明纸牌的牌面值与点数，代码省略 getter 及 setter 方法的定义。

```java
public enum FaceEnum {
    A("A",1),TWO("2",2),THREE("3",3),FOUR("4",4),FIVE("5",5),SIX("6",6)
    ,SEVEN("7",7),EIGHT("8",8),NINE("9",9),TEN("10",10),ELEVEN("J",10)
    ,TWELVE("Q",10),THIRTEEN("K",10);
    private final String name;                    //牌面值
    private final int point;                      //点数
    private FaceEnum(String name,int point) {
        this.name = name;
        this.point = point;
```

                }
            }

3. 纸牌类 Card

封装花色与牌面值两个属性，省略 getter、setter 方法及构造方法。

```java
public class Card {
    private SuitEnum suit;           //纸牌的花色
    private FaceEnum face;           //纸牌的面值
    public String toString() {
        return String.format("%s--%s", suit.getName(),face.getName());
    }
}
```

4. 类 DeckOfCards

设计类 DeckOfCards 描述一副纸牌。封装纸牌数组，提供如下方法。

(1) 一副纸牌的初始化：相当于拿到一副新牌，纸牌按照红桃 A～红桃 K、黑桃 A～黑桃 K、方块 A～方块 K 及梅花 A～梅花 K 的顺序排列。

(2) 查看一副牌：依次输出每张牌的花色及面值。

(3) 洗牌：利用随机数交换纸牌数组中对应位置的纸牌，即随机打乱这副牌。

(4) 发牌：获取并删除纸牌数组中的最后一张牌。

```java
public class DeckOfCards {
    private Card[] cards = new Card[52];
    public void initCards() {        //初始化
        int i = 0;
        for(SuitEnum suit: SuitEnum.values())
            for (FaceEnum face: FaceEnum.values())
                cards[i++] = new Card(suit, face);
    }
    public void shuffle() {          //洗牌
        Card temp = null;
        Random r = new Random();
        for (int i = 0; i < cards.length; i++) {
            int j = r.nextInt(cards.length);
            temp = cards[i];
            cards[i] = cards[j];
            cards[j] = temp;
        }
    }
    public void displayDeck() {      //查看一副牌
        for (int i = 0; i < cards.length; i++) {
            if (i % 13 == 0)
```

```
                System.out.println();
            System.out.print(cards[i] + "\t");
        }
        System.out.println();
    }
    public Card dealCard() {            //发牌
        Card c = cards[cards.length - 1];
        cards = Arrays.copyOf(cards, cards.length - 1);
        return c;
    }
}//类 DeckOfCards 定义结束
```

7.4.2 玩家类

21点游戏分为人-机游戏、人-人游戏两种模式。参与游戏的玩家既可以是电脑玩家，也可以是真人玩家。定义如图7-2所示的玩家类图(省略类属性及方法的定义)。

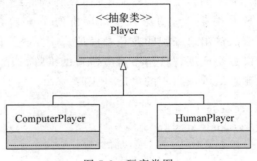

图7-2 玩家类图

1. 抽象类 Player

设计抽象类 Player，描述玩家模板。封装玩家的姓名、玩家手中的牌、总点数及玩家的状态(由枚举类 StateEnum 进行描述)。类 Player 提供如下方法。

(1) 构造方法：初始化玩家状态为"继续"。

(2) 获得一张牌：保存当前扑克牌，计算总点数。如果点数大于21点，则修改玩家状态为"爆牌"，总点数定为-1。

(3) 输出玩家信息：输出玩家姓名、总点数及手中的牌面。

(4) 玩家的比较：根据任务要求，以玩家的总点数确定玩家的排名。

(5) 抽象方法 vote()：描述玩家的玩牌策略，由子类给出具体的实现。

```
public enum StateEnum {
    BUST,STAND,CONTINUE,OVER;
}
public abstract class Player implements Comparable <Player> {
```

```java
    private String name;                //玩家姓名
    private Card[] cards;               //玩家手中的纸牌
    private int totalPoints;            //总点数
    private int cardNum;                //纸牌数目
    private StateEnum state;            //玩家状态
    public Player(String name) {
        this.name = name;
        this.state = StateEnum.CONTINUE;
        cards = new Card[26];
        totalPoints = 0;
        cardNum = 0;
    }
    public void getCard(Card c) {
        cards[cardNum] = c;
        cardNum++;
        totalPoints += c.getPoint();
        if (totalPoints > 21) {
            totalPoints = - 1;
            state = StateEnum.BUST;
        }
    }
    public void display() {
        System.out.printf("%s,点数: %d\n", name,points);
        for(int i= 0;i < cardNum;i++) {
            System.out.print(cards[i]+"\t");
        }
    }
    public int compareTo(Player o) {
        return this.getPoints() - o.getPoints();
    }
    public abstract void vote() throws IOException;
}//抽象类 Player 定义结束
```

2. 电脑玩家子类

以抽象类 Player 为父类生成子类 ComputerPlayer,模拟电脑玩家,实现电脑玩家的游戏策略:点数达到 21 点则结束游戏;点数到达 15 点则停牌。

```java
public class ComputerPlayer extends Player {
    public ComputerPlayer() {
        super("Computer");
    }
    @Override
    public void vote() {
        if(this.getPoints() == 21)
```

```
                this.setState(StateEnum.OVER);
            else if(this.getPoints() >= 15) {
                this.setState(StateEnum.STAND);
            }
            System.out.println("Computer 的状态:"+getState());
        }
    }
```

3. 用户玩家

以抽象类 Player 为父类生成子类 HumanPlayer,描述真实的用户玩家。用户玩家的游戏策略通过用户的选择操作实现。

```
public class HumanPlayer extends Player {
    private Scanner sc = new Scanner(System.in);
    public HumanPlayer(String name) {
        super(name);
    }
    @Override
    public void vote() throws IOException {
        display();
        if(getState() != StateEnum.BUST){
            System.out.println("请选择操作:继续发牌(c) 停牌(s) 结束(o)?");
            String input = sc.nextLine();
            if(input.equals("c"))
                this.setState(StateEnum.CONTINUE);
            else if(input.equals("s")) {
                this.setState(StateEnum.STAND);
                sc.close();
            }
            else if(input.equals("o")) {
                this.setState(StateEnum.OVER);
                sc.close();
            }
        }
        System.out.println(this.getName()+"的状态:"+this.getState());
    }
}
```

7.4.3 游戏类

设计游戏类 BlackJack,封装一副纸牌及参与游戏的多名玩家,提供如下方法。

(1) 构造方法:初始化一副新牌,导入多名游戏玩家。

(2) 游戏主方法 play:首先洗牌,然后为每名玩家分发纸牌,玩家根据手中纸牌的总点数可以选择"停牌""继续"或者"结束"。如果一名玩家已经"爆牌"或者"停牌",则不再继续为该名玩家发牌。最后判断游戏是否结束,游戏结束则输出游戏的赢家。

(3) 判断游戏是否结束 isOver:若满足下列条件之一,则游戏结束。
① 所有玩家选择"结束"。
② 玩家全部"爆牌"。
③ 不存在状态为"继续"的玩家。

(4) 判断并输出赢家:对所有的玩家根据点数和进行降序排序,获取点数最多的玩家。如果仅有一名玩家点数最多,并且其没有"爆牌",则该名玩家胜出;如果有多名玩家点数相同,则为平局。

```java
public class BlackJack {
    private DeckOfCards deck = new DeckOfCards();
    private Player[] players;
    public BlackJack(Player[] p) {
        deck.initCards();
        System.out.println("**********21点游戏**********");
        players = p;
    }
    public void play(){
        deck.shuffle();
        while(true) {
            for(Player p:players) {
                if(p.getState() == StateEnum.CONTINUE) {
                    System.out.println("给"+p.getName()+"发牌");
                    p.getCard(deck.dealCard());
                    try {
                        p.vote();
                    } catch (IOException e) {
                        e.printStackTrace();
                    }
                }
            }
            if(isOver())
                break;
        }
        displayWinner();
    }
    private boolean isOver() {
        int number = 0;
        for(Player p:players) {
            if(p.getState() == StateEnum.OVER)
                return true;
            else if(p.getState() == StateEnum.CONTINUE)
```

```java
                number++;
            }
            if(number < 1)
                return true;
            else
                return false;
        }
        private void displayWinner() {
            Arrays.sort(players,Collections.reverseOrder());
            int i= 1;
            for(; i < players.length;i++) {
                if(players[i].getPoints() != players[0].getPoints())
                    break;
            }
            if(i > 1) {
                System.out.println("平局");
                for(int j= 0;j < i;j++)
                    players[j].display();
            }else {
                System.out.println("优胜者:");
                players[0].display();
            }
        }
}//类 BlackJack 定义结束
```

编写游戏的测试游戏 BlackJackTest。

```java
public class BlackJackTest {
    public static void main(String[] args) {
        BlackJack bj = new BlackJack(new Player[]{new ComputerPlayer()
                            , new HumanPlayer("我")});
        bj.play();
    }
}
```

7.5 练一练

1. 随机生成字符串数组,其中,随机字符仅能由英文字母组成,并且每个字符串的长度相等。编程统计随机生成的字符串数组中包含多少种模式以及每种模式出现的次数。根据程序注释,在横线处完成代码。

注意:一种模式是指由连续相同的若干字母组成,如字符串"aabbcddaa"中存在 4 种模式,分别为"aa""bb""c"及"dd",4 种模式出现的次数分别为 2、1、1、1。

```
class Model{                    //定义模式类 Model
```

```java
    private String modelName;
    private int count;      //模式出现的次数
    public String toString() {
        return String.format("model:% s count:% d",
            getModelName(),this.getCount());
    }
}
public class StringRandom {       //定义类 StringRandom 封装随机字符串数组及其操作
    private String[] strs;
    /*实现长度为 stringCount 的随机字符串数组的初始化,每个字符串的长度限定为
stringLength*/
    public StringRandom(intstringCount,intstringLength) {
        strs = new String[stringCount];
        StringBuilder sb = new StringBuilder();
        for(int i = 0;i < strs.length;i++) {
            _____//清空 sb 的内容
            for(int j = 0; j < stringLength; j++){
                _____//将随机生成的字符追加到 sb 中
            }
            strs[i] = sb.toString();
        }
    }
    public void display() {
        for(String str:strs)
            System.out.println(str);
    }
    private char getRandomChar(char start,char end) {
        _____//随机生成字符,要求字符取值范围定义为[start,end]
    }
    public Model[] getModel() {   //统计随机字符串数组中出现的模式
        Model[] models = new Model[strs.length];
        int modelCount = 0;
        _____//复制 strs,生成字符串数组 strTemp
        for(int i = 0; i < strTemp.length; ) {//对 strTemp 进行排序
            int count = 1;
            int j = i+1;
            for(; j < strTemp.length; j++)
                if(strTemp[i].equals(strTemp[j]))
                    count++;
                else
                    break;
            models[modelCount++] = new Model(strTemp[i],count);
            i = j;
        }
```

```
            return Arrays.copyOf(models,modelCount);
        }
        public static void main(String[] args) {    //功能测试
            StringRandom sr = new StringRandom(100,2);
            sr.display();
            for(Model m:sr.getModel())
                System.out.println(m);
        }
    }
```

2. 随机生成日期数组，其中，日期限定在 2001—2003 年之间，要求对随机数组进行排序，并按照"年/月/日 时:分:秒"的格式输出。

第 8 章 正则表达式与异常处理
——个人通讯录（三）

8.1 个人通讯录（三）

设计个人通讯录（三），重新定义联系人 Contract 类，需要保证联系人信息的有效性。联系人信息必须满足以下条件：
(1) 姓名不能为空；
(2) 性别可以为空、"男"或"女"；
(3) 电子邮箱可以为空或者满足电子邮箱格式规范；
(4) 联系电话不能为空，可以包含固定电话或者移动电话号码。

8.2 程序设计思路

联系人 Contract 类包含姓名（name）、性别（gender）、电子邮件（email）、若干个联系电话（phones）等成员变量，为每个成员变量生成 getter 及 setter 方法，在 setter 方法中对输入参数的样式进行判定，对于不满足样式要求的数据，通过异常处理机制对其进行响应。

对字符串的格式进行判定，需要引入正则表达式。正则表达式是对字符串操作的一种规则逻辑，采用事先定义好的一些特定字符及特定字符的组合构成一个"规则字符串"，利用这个"规则字符串"表达对字符串的一种匹配和过滤逻辑。

8.3 关键技术

8.3.1 正则表达式

Java 4 增加了对正则表达式的支持。正则表达式又称规则表达式，是处理字符串的强大工具，可以用于字符串的匹配、查找与替换。例如下面的代码使用了正则表达式".*abc.*"，用来判定字符串中是否包含"abc"子串。

```
String reg = ".*abc.*";
boolean isMatched = Pattern.matches(reg, "eeeabcffff");    //isMatched= true
```

其中，字符".*"匹配除换行符之外的任意个任意字符；java.util.regex.Pattern 类中定义的静态方法 matches(String regx, String str)用于判定字符串 str 与正则表达式 regx 是否匹配。

1. 正则表达式

正则表达式定义了字符串的模式,用于匹配指定格式的字符串。

(1) 单个字符的匹配。

表 8-1 列出了常用于匹配单个字符的正则表达式。

表 8-1 单个字符的匹配

字 符	含 义
x	字符 x(x 可以表示任何一个确定的字符)
.	任意一个字符
\\	匹配"\"字符
\t	匹配制表符 tab
\n	匹配换行符
\r	匹配回车符
\uhhhh	十六进制 0xhhhh 表示的 Unicode 字符
\\d	0~9 中任意一个数字
\\D	任意一个非数字字符
\\s	空白字符,包括空格、制表符、回车符、换页符、换行符等
\\S	非空白字符
\\w	任意单词字符,包括数字 0~9、26 个英文字母和下画线
\\W	所有非单词字符
\\p{Lower}	任意小写字母
\\p{Upper}	任意大写字母
\\p{ASCII}	任意 ASCII 码字符
\\p{Alpha}	字母
\\p{Digit}	数字 0~9
\\p{Alnum}	字母和数字
\\p{Punct}	标点符号,包括!"#$%&'()*+,-./:;<=>?@[\]^_`{|}~

(2) 方括号表达式。

正则表达式中可以使用方括号表达式提供更灵活的匹配方式。表 8-2 列了出方括号表达式的形式及含义。

(3) 边界匹配。

Java 正则表达式还提供了如表 8-3 所示的边界匹配符。

(4) 多字符的匹配。

在正则表达式中引入数量字符,可以匹配多个字符。多字符的匹配如表 8-4 所示。

表 8-2　方括号表达式

方括号表达式	含　义
[…]	匹配"[]"中包含的任意一个字符。如[abc]，匹配字符 a、b 或者 c 中的任意一个
[^…]	匹配除了"[]"中包含的字符以外的任意字符。如[^abc]，匹配除了字符 a、b 或者 c 之外的任意字符
[a-z]	匹配 a~z 中的任意一个字符。如[a-d]，匹配 a、b、c、d 中的任意一个字符，与[acbd]相当
[a-z&&[12xyz]]	匹配同时满足[a-z]及[12xyz]两个条件的字符，即匹配 x、y 或 z 中的任意一个字符

表 8-3　边界匹配符

边界匹配符	含　义
^	匹配字符串的开始或者行的开始（针对多行字符串）。如^a 匹配"ab"，但是不能匹配"ba"
$	匹配字符串的结束或者行的结束（针对多行字符串）。如 a$ 匹配"ba"，但是不能匹配"ab"
\\b	匹配单词边界，即字与空格间的位置。如 ab\\b 匹配"aab"中的"ab"，但不匹配"aabb"中的"ab"
\\B	非单词边界。如 ab\\B 匹配"aaba"中的"ab"，但不匹配"aab"中的"ab"
\\A	匹配字符串的开始
\\Z	匹配字符串的结束或字符串结尾的 \n 之前
\\z	匹配字符串的结束

表 8-4　多字符的匹配

多字符匹配	含　义
x{n}	字符 x 重复 n 次
x{n,}	字符 x 至少重复 n 次
x{n,m}	字符 x 至少重复 n 次，至多重复 m 次。如："o{1,3}"匹配"toood"中的 3 个 o
x?	字符 x 出现 1 次或者 0 次
x*	字符 x 出现多次或者 0 次
x+	字符 x 至少出现 1 次

2．正则表达式的应用

Java 中与正则表达式相关的工具主要在 java.util.regex 包中，此包提供 Pattern 类，用于创建一个正则表达式，也可以说创建一个规则模式。Pattern 对象是正则表达式编译之后在内存中的表示形式。

(1) 编译正则表达式。

```
public static Pattern compile(String regex)
```

Pattern 类的构造方法是私有的,所以不可以直接通过构造方法创建 Pattern 对象。通常通过 Pattern 类的静态方法 compile()编译正则表达式 regex,构建 Pattern 对象。

(2) Pattern 类常用方法。

```
public static boolean matches(String regex, CharSequence input)
```

静态方法 matches()根据正则表达式 regex 生成匿名的 Pattern 对象,判定字符序列 input 与匿名 Pattern 对象是否匹配。CharSequence 是一个字符序列接口,常见的实现类包括 String、StringBuffer、StringBuilder 等,如:

```
Pattern.matches("\d+", "1234");       //匹配一个或者多个数字,返回 true
```

8.3.2 异常概述

Java 提供异常机制从而识别及响应程序在编译及运行期间产生的错误,有效的异常处理能使程序更加健壮且易于调试。常见的异常类型及其继承关系如图 8-1 所示。

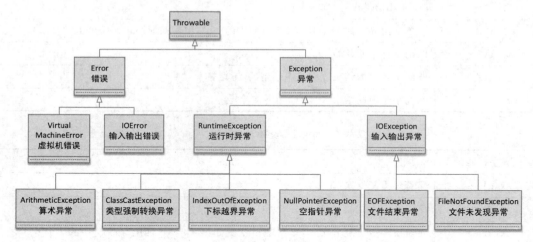

图 8-1 常见的异常类型及其继承关系

Java 的所有异常类都是从 Throwable 继承而来的,扩展 Throwable 生成 Error 和 Exception 子类,其中,Error 表示故障与虚拟机相关,如 Java 虚拟机运行错误(VirtualMachineError)、输入/输出错误(IOError)等。这些错误是应用程序无法捕获或处理的,因为它们超越了应用程序的控制和处理能力。

Exception 异常子类是应用程序本身可以处理的异常。Exception 异常通常分为 Checked 异常和 Runtime 异常两大类。Checked 异常称为可检查的异常,它是需要在应用程序中显式处理的异常,否则程序将无法通过编译。Checked 异常的发生是可以预测的,也是可以合理处理的。例如 IOException,当打开一个文件时,可能会发生文件不存在的异常,所以在应用程序中需要对 IOException 进行处理。Exception 的子类中除了 RuntimeException 及其子类以外,都属于 Checked 异常。另一类就是 Runtime 异常,包括 RuntimeException 及其子类。Runtime 异常在应用程序中既可以显式捕获并处理,也可以不处理,编译时不检查也不会出现编译错误,所以 Java 中 Runtime 异常的处理更加灵活。

8.3.3 异常处理

1. 异常处理

异常处理的一般形式如下。

```
try {
        //可能发生异常的代码序列
    }
[catch (异常类 1 e1) {
        异常处理代码序列 1;
    } catch (异常类 2 e2) {
        异常处理代码序列 2;
    }
...]
[finally {
        //无论是否发生异常都会执行的代码序列
    }]
```

程序执行 try 代码块时如果发生异常,则系统会自动生成一个异常对象,并且将该异常对象交由与之匹配的 catch 子句进行处理;如果没有发生异常,则所有的 catch 子句都不执行。finally 子句位于 catch 子句的后面,通常用于释放 try 子句中打开的物理资源。无论是否发生异常,finally 子句中的代码都会执行。

在异常处理的 3 个子句中,只有 try 子句是必需的,catch 子句和 finally 子句都是可选项,但是 catch 子句和 finally 子句必须至少出现一项。如果有多个 catch 子句同时出现,那么父类异常处理的子句必须位于子类异常处理的子句之后。例 8-1 示范了异常处理语句的使用。

例 8-1

```
public class ExceptionTest {
    public static void main(String[] args) {
        String filename = "c:/exceptionTest.txt";
        File file = new File(filename);
        ObjectInputStream ois = null;
        try {
            ois = new ObjectInputStream(new FileInputStream(file));
            Object obj = ois.readObject();
```

```
            System.out.println(obj);
        } catch (FileNotFoundException e) {
            System.out.println("文件未找到");
        } catch (EOFException e) {
            System.out.println("文件结束");
        }catch (IOException e) {
            System.out.println("读文件时发生输入输出异常");
        } catch (ClassNotFoundException e) {
            System.out.println("类未找到");
        }finally {
            if(ois != null)
                try {
                    ois.close();
                } catch (IOException e) {
                    System.out.println("文件关闭时发生输入输出异常");
                }
        }
    }
}
```

例 8-1 将从文件中读取一个对象输出到控制台的代码放在了 try 子句中,在读取对象的过程中,系统可能会抛出如下异常对象。

(1) FileNotFoundException:当不存在 c:\exceptionTest.txt 文件时抛出。
(2) EOFException:读取对象遇到文件结束时抛出。
(3) IOException:读文件时发生其他输入输出异常时抛出。
(4) ClassNotFoundException:读文件时未找到 Object 对象时抛出。

抛出的异常对象会交由相匹配的 catch 子句执行。最后,无论是否发生异常,都会执行 finally 子句,关闭打开的文件,释放资源。

2. 异常的增强处理

从 Java 7 开始,Java 在异常处理机制上做了一些增强处理,主要体现在多异常捕获与自动关闭资源。

所谓多异常捕获,就是定义一个 catch 子句以捕获多种类型异常对象,例如:

```
try {
    ois = new ObjectInputStream(new FileInputStream(file));
    Object obj = ois.readObject();
    System.out.println(obj);
}
catch(FileNotFoundException | ClassNotFoundException e) {          ①
    System.out.println("文件未找到或者类未发现.");
} catch (EOFException e) {
    System.out.println("文件结束.");
```

```
}catch (IOException e) {
    System.out.println("读文件时发生输入输出异常.");
}
```

程序中代码①处的 catch 子句可以捕获两种类型的异常对象。需要注意的是，在一个 catch 子句中声明捕获的这些异常类中，不能出现重复的类型，也不允许其中的某个异常是另一个异常的子类，否则会出现编译错误，如：

```
try {
...
}
catch(FileNotFoundException | IOException | ClassNotFoundException e){    ②
}
```

代码②处会产生编译错误。

异常处理经常会遇到与例 8-1 类似的 finally 子句，程序会显得很不简洁。为了解决这个问题，Java 7 之后推出了 try-with-resources 声明以替代之前的 try-finally 声明。try-with-resources 是定义了一个或多个资源的 try 声明，try 语句在该语句结束时自动关闭这些资源。try-with-resources 可以确保每一个资源在处理完成后都会被关闭。这些资源必须实现 AutoCloseable 接口或者 Closeable 接口。下面的程序使用 try-with-resources 自动关闭资源。

```
public class TryWithResourcesTest {
    public static void main(String[] args) {
        String filename = "c:/exceptionTest.txt";
        File file = new File(filename);
        try (ObjectInputStream ois= new ObjectInputStream(
                                new FileInputStream(file))){           ①
            Object obj = ois.readObject();
            System.out.println(obj);
        } catch(FileNotFoundException | ClassNotFoundException e) {
            System.out.println("文件未找到或者类未发现.");
        } catch (EOFException e) {
            System.out.println("文件结束.");
        }catch (IOException e) {
            System.out.println("读文件时发生输入输出异常.");
        }
    }
}
```

上述代码行①处声明并初始化了输入流对象 ois，执行 try-catch 语句之后将自动关闭输入流对象 ois。

3. 抛出异常

如果一个方法可能会产生异常，则在方法中除了采用 try-catch 语句进行捕获处理之外,还可以交由方法的上一级调用者处理,这时该方法需要使用throws关键字声明抛出异常。throws关键字放在方法签名的尾部,语法格式为

［修饰符］返回类型　方法名(参数列表) throws 异常类型1,异常类型2,…
{
　　//方法体
}

例8-2示范了throws声明抛出异常。

例8-2

```java
public class ThrowsTest {
    public Object readFromFile(String filename)
                throws ClassNotFoundException, FileNotFoundException
                        , IOException{                                    ①
        File file = new File(filename);
        ObjectInputStream ois= new ObjectInputStream(
                        new FileInputStream(file));
        Object obj = ois.readObject();
        return obj;
    }
    public static void main(String[] args) {
        ThrowsTest test = new ThrowsTest();
        try {
            Object obj = test.readFromFile("c:/exceptionTest.txt");
            System.out.println(obj);
        } catch (ClassNotFoundException | IOException e) {
            System.out.println("读文件错误");
        }
    }
}
```

例8-2的代码行①采用throws子句声明了该方法将抛出异常,在readFromFile()方法体中没有异常的捕获及处理,main()方法调用了声明抛出异常的readFromFile()方法,这时就需要在main()中对异常进行处理。main()方法可以使用try-catch语句对异常进行捕获,并处理如例8-2所示;也可以进一步使用throws声明抛出异常,main()方法抛出的异常将由JVM进行处理。

如果父类中定义了使用throws声明抛出异常的方法,那么在子类中重写该方法时需要注意以下问题:

(1) 如果父类方法抛出单个异常,那么子类的重写方法,只能抛出父类的异常或者该异常的子类。

(2) 如果父类方法抛出多个异常,那么子类在重写该方法时只能抛出父类异常的子集。

(3) 如果父类或者接口的方法中没有异常抛出,那么子类在重写方法时也不可以抛出异常。

前面介绍的异常都是系统自动抛出的异常,在应用程序中,还可以根据业务逻辑的需要自行抛出异常。其语法格式为

throw <Throwable 类及其子类对象>;

注意:throw 后面不能跟其他语句,因为 throw 后面的语句将无法执行。例如下面的代码所示。

```
public Object readFromFile(String filename) throws Exception {
    File file = new File(filename);
    try (ObjectInputStream ois= new ObjectInputStream(
                            new FileInputStream(file))){
        Object obj = ois.readObject();
        if(!check(obj))                                         ①
            throw new Exception("数据不合法.");                    ②
        else
            return obj;
    } catch (ClassNotFoundException | IOException e) {
        return null;                                            ③
    }
}
```

代码行①调用方法 check()判断获取的对象 obj 是否符合业务要求(这里省略 check()方法的定义),若 check()方法返回 false,则利用 throw 语句抛出异常,如代码行②;代码行③对文件读写过程系统可能抛出的异常进行处理并返回 null 对象。

4. 获取异常信息

应用程序通常需要在异常处理的 catch 子句中获取相关的异常信息,可以使用 Exception 类提供的如下方法获取。

(1) public String getMessage()。

获取异常的描述字符串。

(2) public void printStackTrace()。

输出异常的跟踪栈信息。

8.3.4 自定义异常类

当 Java 预定义的类库不能满足用户需要时,便可以自定义需要的异常类。自定义异常类必须继承于已有的异常类,即用户自定义的异常类都必须直接或间接的是 Exception 类的子类。以下程序创建了一个自定义异常类。

```
class IllegalDataException extends Exception{
```

```java
    public IllegalDataException() {
    }
    public IllegalDataException(String str) {
        super(str);
    }
}
```

上述程序声明了异常子类 IllegalDataException，其中定义了两个构造方法，实现了 IllegalDataException 实例的初始化。自定义异常子类相较异常父类 Exception 而言能更准确地描述异常，有利于异常情况的区分。

8.4 个人通讯录(三)设计步骤

8.4.1

8.4.1 自定义异常子类

针对姓名、性别、电子邮件、电话分别定义格式异常子类。

```java
class NameException extends Exception{
    public NameException() {
        super("姓名为空");
    }
}
class GenderException extends Exception{
    public GenderException() {
        super("性别格式错误");
    }
}
class EmailException extends Exception{
    public EmailException() {
        super("邮箱格式错误");
    }
}
class PhoneException extends Exception{
    public PhoneException() {
        super("电话号码格式错误");
    }
}
```

8.4.2

8.4.2 Contract 类

在个人通讯录(二)的联系人 Contract 类的基础上重写各个实例变量的 setter 方法。

1. setName 方法

```
public void setName(String name) throws NameException{
    if(name == null || name.equals(""))
        throw new NameException();
    this.name = name;
}
```

2. setGender 方法

```
public void setGender(String gender) throws GenderException{
if(gender == null || gender.equals("男") || gender.equals("女") || gender.equals(""))
        this.gender = gender;
    else
        throw new GenderException();
}
```

3. setEmail 方法

```
public void setEmail(String email) throws EmailException{      //email 的正则表达
    String regex = "^[a-zA-Z0-9_-]+@[a-zA-Z0-9_-]+(\.[a-zA-Z0-9_-]+)+$";
    if(email == null||email.equals("") || Pattern.matches(regex, email))
            this.email = email;
    else
        throw new EmailException();
}
```

4. setPhones 方法

```
public void setPhones(String[] phones) throws PhoneException{
    if(phones == null || phones.length == 0)
        throw new PhoneException();
    String telReg = "^(0[1-9]\d{1,2}\-)?\d{7,8}$ ";          //固定电话的正则表达式
    String phoneReg = "^1[35789][0-9]{9}$ ";                 //移动电话的正则表达式
    for(String phone:phones)
        if(phone != null && (Pattern.matches(telReg, phone)
                    || Pattern.matches(phoneReg, phone)))
            this.phones = phones;
        else
            throw new PhoneException();
}
```

8.5 练一练

1. 运行下列代码,分析运行结果。

```java
public class ExceptionTest{
    public static void main(String args[]) {
        int i = 0;
        int a[] = { 1, 2, 3, 4, 5 };
        for (i = 0; i < 6; i++) {
            try {
                System.out.print("a[" + i + "]/" + i + "= " + (a[i]/i)+" ");
            } catch (ArrayIndexOutOfBoundsException e) {
                System.out.print("数组下标越界异常!");
            } catch (ArithmeticException e) {
                System.out.print("算术异常!");
            } catch (Exception e) {
                System.out.print("捕获" + e.getMessage() + "异常!");
            } finally {
                System.out.println("i= " + i);
            }
        }
    }
}
```

2. 在 try_catch_finally 的异常处理机制中,如果 catch 中包含 return 语句,那么是否还会执行 finally 对应的语句序列?

3. 自定义异常的作用是什么?

第 9 章 集合——个人通讯录（四）

9.1 个人通讯录（四）

设计个人通讯录（四），要求：
（1）在通讯录系统中使用集合替代数组，重新构建通讯录，实现通讯录的增、删、改、查等操作；
（2）使用外部比较器或者 Comaprable 接口实现联系人对象的比较，在此基础上按照姓名拼音升序排序。

9.2 程序设计思路

个人通讯录（一）至个人通讯录（三）已经实现了通讯录的基本功能，其中对若干联系人的存储采用了数组这种数据结构。数组定义时需要事先确定要保存的对象数量，并且这个数组的长度一旦确定就不可变了。如果应用程序需要存储一系列可以动态增长的数据，那么数组就会显得很不灵活。

Java 提供了大量性能优秀的集合类，用于存储数量不确定的数据对象，并提供对集合元素的基本操作，如增、删、改、查等。由于集合框架中的接口与实现类众多，存储特性与性能各不相同，所以在应用程序中要注意甄别，应根据程序的功能场景和性能需求选用最合适的集合类。

9.3 关键技术

9.3.1 集合概述

Java 中能够容纳多个数据对象的数据结构称为容器。数组是最常见的一种容器，数组的长度是固定且不可改变，采用数组存储动态变化的数据时会显得不太便捷。同样作为数据的容器，Java 集合类主要用于不确定数量的对象存储，并提供栈、队列等常用的数据结构。除此之外，Java 集合还可用于保存具有映射关系的关联数据。

集合相关接口、类及继承关系分别如图 9-1 及图 9-2 所示。

Java 集合框架主要包括两种类型的容器：一种是派生自 java.util.Collection 接口的集合，用来存储若干数据对象的引用，Collection 接口派生 3 个子接口，分别是 List、Set 和 Queue。图 9-1 中常用的实现类包括 HashSet、TreeSet、ArrayList 及 LinkedList。另一种集合类派生自 Map 接口，存储键-值对映射。Map 接口的常用实现类包括 HashMap 和

Hashtable,如图 9-2 所示。

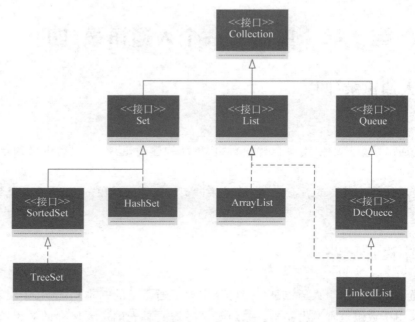

图 9-1　接口 Collection 及其子接口与实现类

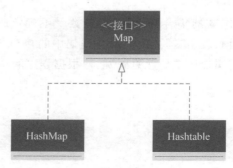

图 9-2　接口 Map 及其实现类

9.3.2　Collection 接口

　　Java.util.Collection 是最基本的集合接口,作为容器类型,Collection 接口用来存储一组对象的引用,不可以存储基本类型数据。Collection 接口提供了大量的方法,可以实现集合元素的添加、删除及访问。

```
public boolean add(Object e)
```

　　向集合添加元素 e,添加成功则返回 true,否则返回 false。

```
public boolean addAll(Collection c)
```

将集合 c 中的所有元素添加到当前集合。

```
public boolean remove(Object o)
```

删除集合中的指定元素 o,若当前集合中包含多个元素 o,则仅删除第一个元素,删除成功则返回 true,否则返回 false。

```
public boolean removeAll(Collection c)
```

从当前集合中删除集合 c 中的所有元素;删除成功则返回 true,否则返回 false。

```
public boolean contains(Object o)
```

查询当前集合是否包含指定元素 o;查询成功则返回 true,否则返回 false。

```
public boolean containsAll(Collection c)
```

检测当前集合是否包含集合 c 的所有元素。

```
public int size()
```

返回集合中元素的个数。

例 9-1 示范了 Collection 接口的方法应用。

例 9-1

```java
class Student{          //定义类 Student
    private int no;
    private String name;
    public String toString() {
        return name;
    }
}
```

定义集合测试类 CollectionTest。

```java
public class CollectionTest {
    public static void main(String[] args) {
        Collection c1 = new ArrayList();
        c1.add(new Student(1,"李白"));        //添加元素
        c1.add(new Student(2,"杜甫"));
        System.out.println("c1 中元素为:");
        System.out.println(c1);
        Collection c2 = new ArrayList();
        c2.add(new Student(3,"苏轼"));
        c2.addAll(c1);                        //将集合 c1 中的所有元素添加到 c2 中
        System.out.println("c2 中元素为:");
        System.out.println(c2);
```

```
            c1.remove(new Student(1,"李白"));                    ①
            System.out.println("c1 中元素为：");
            System.out.println(c1);
            System.out.println(c1.contains(new Student(2,"苏辙")));②
        }
    }
```

例 9-1 首先定义了实体类 Student，用来描述学生基本信息。测试类 CollectionTest 中创建了两个集合实例 c1 和 c2，用来示范 Collection 接口中方法的使用。代码行①从集合 c1 中删除了指定元素 new Student(1,"李白")，运行上述代码时会发现删除操作失败。分析程序可以发现，当集合 c1 查找指定元素 new Student(1,"李白")时，会调用对象的 equals 方法以匹配待删除元素。默认情况下，equals 方法仅在两个对象引用相等时才返回 true，所以需要在 Student 类中重写 equals 方法，代码如下。

```
public boolean equals(Object obj) {
    if(this == obj)
        return true;
    if(obj != null && obj instanceof Student) {
        return this.no == ((Student)obj).getNo();
    }else
        return false;
}
```

同理，代码行②中的 contains 方法也会调用对象的 equals()方法以匹配待查询元素。

9.3.3 集合的遍历

Collection 集合通常采用以下两种方式遍历集合元素。

1. Iterator 接口

Java 系统提供了迭代器 Iterator，用于遍历 Collection 集合中的元素。Iterator 接口中定义了以下 3 个方法。

`public boolean hasNext()`

检测是否存在还没有被遍历的元素，存在则返回 true，否则返回 false。

`public Object next()`

返回下一个迭代元素。

`public void remove()`

删除迭代器对应的集合中上一次 next()方法返回的元素。

例 9-2 示范了利用 Iterator 接口遍历集合元素。

例 9-2

```java
public class IteratorTest {
    public static void main(String[] args) {
        Collection c = new ArrayList();
        c.add(new Student(1,"李白"));
        c.add(new Student(2,"杜甫"));
        c.add(new Student(3,"苏轼"));
        Iterator itor = c.iterator();           //获取迭代器对象
        while(itor.hasNext()) {
            Student s = (Student)(itor.next()); //获取下一个集合元素
            System.out.println(s);
            if(s.getNo() == 1)
                itor.remove();                  //删除上次next()返回的集合元素
                //c.remove(s);                                                ①
        }
        System.out.println(c);
    }
}
```

注意：在使用 Iterator 迭代器对象遍历集合元素时，不能通过集合变量修改集合元素，如例 9-2 中的代码行①，否则系统会抛出 java.util.ConcurrentModificationException 异常。

2. foreach 循环

Java 提供了便捷的 foreach 循环语句迭代集合元素。例如以下代码利用 foreach 循环遍历了集合元素。

```java
for(Object o:c) {
    Student s = (Student)(o);
    System.out.println(s);
    /*if(s.getNo() == 1)
        c.add(new Student(4,"王维"));*/                                        ①
}
```

利用 foreach 循环遍历集合元素与利用 Iterator 迭代器遍历集合元素相似，也不能通过集合变量修改集合元素，如代码行①，否则系统会抛出 java.util.ConcurrentModificationException 异常。

9.3.4 Set 接口及其实现类

Set 接口及其实现类的共同特点：与数学中的集合概念相吻合，即 Set 集合中不允许包含重复元素；另外，Set 集合不能记住元素的添加顺序。HashSet、LinkedHashSet 与

TreeSet 是 Set 接口的主要实现类。作为 Collection 的子接口，Set 接口继承了其父接口中定义的方法。

1. HashSet

HashSet 是 Set 接口的典型实现。HashSet 按哈希算法存储集合中的元素，因此具有良好的查找性能。HashSet 中元素的排列顺序可能与元素的添加顺序不同；另外，HashSet 集合的元素值允许为 null。

注意：存放在 HashSet 中的类对象需要同时重新定义 equals 方法和 hashCode 方法，以保证参与比较的两个对象之间的 equals() 方法与 hashCode 方法的返回值是一致的，即若两个对象通过 equals() 方法进行比较返回了 true，则这两个对象应该具有相同的 hashCode 值。

例 9-3 示范了 HashSet 类的应用。

例 9-3　定义类 Student。

```java
class Student{
    private int no;
    private String name;
    public String toString() {
        return name;
    }
    public boolean equals(Object obj) {
        if(this == obj)
            return true;
        if(obj != null && obj instanceof Student) {
            return this.no == ((Student)obj).getNo();
        }else
            return false;
    }
    public int hashCode() {
        return this.no;
    }
}
```

定义测试类 HashSetTest。

```java
public class HashSetTest {
    public static void main(String[] args) {
        Set s = new HashSet();
        s.add(new Student(2,"杜甫"));
        s.add(new Student(1,"李白"));
        s.add(null);            //添加 null 元素
        System.out.print("添加元素的结果为：");
        System.out.println(s.add(new Student(2,"苏轼")));    ①
```

```
        System.out.print("添加操作后,集合 s 中元素为: ");
        System.out.println(s);                                              ②
        s.remove(new Student(2,""));        //移除集合元素              ③
        System.out.print("移除操作后,集合 s 中元素为: ");
        System.out.println(s);
    }
}
```

例 9-3 首先定义了学生类 Student,重写了 equals()方法,以学生的编号作为 equals()方法进行比较的依据。同时,重写了 hashCode()方法,以保证 equals()与 hashCode()这两个方法返回结果的一致性。

HashSetTest 测试程序向集合 s 添加包含 null 元素在内的 3 个元素,代码行①向集合 s 添加对象 new Student(2,"苏轼"),该对象与集合中已存在的元素 new Student(2,"杜甫")具有相同的 hashCode 值,并且这两个对象通过 equals()进行比较后的返回值为 true,系统认定这两个对象是相同元素,所以添加操作失败,add()方法返回 false。另外,代码行②输出了集合的所有元素,其排列位置与添加的顺序不同。最后,代码行③处的移除操作通过 equals()方法逐一将集合元素与对象 new Student(2,"")进行比较,最终找到并移除了集合元素 new Student(2,"杜甫")。

2. LinkedHashSet

HashSet 的子类 LinkedHashSet 使用链表维护元素的插入顺序,也就是说,当遍历 LinkedHashSet 集合中的元素时,将会按元素的添加顺序访问集合中的元素。因为链表的引入,LinkedHashSet 类的插入与删除操作性能略低于 HashSet,但是加快了遍历操作的速度。

3. TreeSet

TreeSet 直接实现了接口 SortedSet,可以确保加入集合的元素处于自然排序状态。所谓自然排序,是指 TreeSet 会调用集合元素的 compareTo(Object o)方法以比较元素之间的大小关系,然后将集合中的元素按升序排列。

集合 TreeSet 要求加入集合的实体类必须实现 java.lang.Comparable 接口,否则系统会抛出 java.lang.ClassCastException 异常。实现 Comparable 接口需要重写 int compareTo(Object o)方法,该方法用于比较两个对象的大小。若 obj1.compareTo(obj2)方法返回 0,则表示这两个对象相等;若返回一个正整数,则表示 obj1 大于 obj2;若返回一个负整数,则表示 obj1 小于 obj2。

例 9-4 示范了 TreeSet 类的常用方法。

例 9-4 重构例 9-3 中 Student 类的定义。

```
Public class Student implements Comparable{
    //省略相同部分代码
    @Override
```

```java
    public int compareTo(Object o) {
        Student s = (Student)o;
        return this.getNo()-s.getNo();
    }
}
```

Student 类实现 Comparable 接口重写 int compareTo(Object o)方法,定义 Student 对象的比较行为,判定编号小的 Student 对象小于编号大的 Student 对象。

定义测试类 TreeSetTest。

```java
public class TreeSetTest {
    public static void main(String[] args) {
        TreeSet s = new TreeSet();
        s.add(new Student(3,"苏轼"));
        s.add(new Student(2,"杜甫"));
        s.add(new Student(1,"李白"));
        System.out.println("集合的所有元素: "+s);              ①
        //s.add(null);                                        ②
        System.out.print("添加相同编号的元素: ");
        System.out.println(s.add(new Student(2,"苏轼")));    ③
    }
}
```

例 9-4 采用 TreeSet 类定义集合类型。代码行①输出按照编号升序排列的集合元素。代码行②向 TreeSet 集合插入 null 元素,系统抛出 NullPointerException 异常,验证了 TreeSet 集合与 HashSet、LinkedHashSet 不同,不允许插入 null 元素。代码行③插入相同编号的 Student 实例失败,add()方法返回 false。TreeSet 类作为 Set 接口的实现类,遵循 Set 集合的一般规则:集合中不允许包含相同的元素。

9.3.5 List 接口及其实现类

List 集合中的每个元素都按元素的添加顺序设置顺序索引。List 集合允许出现重复元素。List 作为 Collection 接口的子接口,除了继承父接口方法之外,还提供如下操作集合的方法:

public Object get(int index)

返回集合中指定索引的元素。

public Object set(int index, Object element)

用 element 元素替换集合中 index 处的元素。

public void add(int index, Object element)

在集合的指定位置插入元素 element。

```
public Object remove(int index)
```
移除集合中指定位置的元素。

```
public int indexOf(Object o)
```
返回集合中首次出现元素 o 的索引,如果集合不包含此元素,则返回-1。

List 接口的常用实现类是 ArrayList 和 Vector。ArrayList 和 Vector 类的使用方法基本相同,两者的主要区别在于 ArrayList 类是线程不安全的,也就是当多个线程操作同一个 ArrayList 集合时,如果存在多个线程修改集合元素,则系统将无法保证该集合的同步性,需要通过程序控制集合的同步;而 Vector 类是线程安全的(关于线程的具体内容请参见第 13 章)。正因为 Vector 类是线程安全的,所以 Vector 类的性能要低于 ArrayList 类。另外,由于历史原因,Vector 类除了实现 List 接口中定义的方法之外,还存在一些功能重复的方法,这会造成 Vector 类的结构不清晰,所以在应用中通常不推荐使用 Vector 类。

9.3.6 Map 接口及其实现类

Map 用于保存具有映射关系<key,value>的数据。key 和 value 之间存在单向一对一关系,即通过指定的 key 总能找到唯一确定的 value。Map 中的 key 不允许重复,key 的对象必须同时重写 hashCode()方法及 equals()方法。

Map 接口的常用方法如下。

```
public Object put(Object key, Object value)
```
将新的映射<key,value>添加到当前 Map。返回之前与 key 关联的 value 值,如果之前没有关联值,则返回 null。

```
public Object remove(Object k)
```
从当前映射中移除以 k 为键的映射关系,返回 k 对应的 value。

```
public Object get(Object k)
```
获取映射中与 k 相对应的 value。

```
public boolean containsKey(Object k)
```
判断映射中是否含有键 k。

```
public boolean containsValue(Object value)
```
判断映射中是否存在值 value。

```
public Set keySet()
```
获取当前映射的所有键。

```
public Collection values()
```

获取当前映射的所有值。

HashMap 和 Hashtable 都是 Map 接口的典型实现类。Hashtable 是线程安全的,而且不允许使用 null 作为 Hashtable 集合的键或值;HashMap 是线程不安全的,它可以使用 null 作为集合的键或值。例 9-5 示范了 HashMap 的一般用法。

例 9-5

```java
public class HashMapTest {
    public static void main(String[] args) {
        Map map = new HashMap();
        Student s1= new Student(1,"李白");
        map.put(s1.getNo(), s1);
        Student s2= new Student(2,"杜甫");
        map.put(s2.getNo(), s2);
        System.out.println(map.put(2, new Student(2,"苏轼")));       ①
        System.out.println(map.containsValue(new Student(2,"杜甫"))); ②
        for(Object key:map.keySet())
            System.out.println(map.get(key));
    }
}
```

上述代码省略了 Student 类的定义,具体代码见例 9-4。例 9-5 首先通过 put()方法向映射 map 添加两个＜key,value＞对。代码行①再次添加 key＝2 的元素,新的 Student 实例覆盖并返回原有的 value,所以代码行①输出"杜甫"。代码行②的 containsValue()方法判断映射的 value 集合中是否存在 new Student(2,"杜甫")实例对象。判断是否存在的标准是两个 Student 对象通过 equals()进行比较后是否返回 true。例 9-4 中重写了 Student 类的 equals()方法,若两个 Student 实例的编号相等,则 equals()方法返回 true。所以,代码行②判定映射的 value 集合中包含 new Student(2,"杜甫")实例对象,输出 true。例 9-5 的最后一段代码使用 for 语句遍历了 map 映射。

9.3.7 泛型

集合作为数据的容器对元素的类型是没有限制的,被置于集合中的元素经过编译之后都会泛化为 Object 类型,因此从集合中取出的元素通常需要经过强制类型转换才能使用其特有的方法,而使用不当则会引发 ClassCastException 异常。

自 Java 5 之后,Java 引入了泛型的概念,允许在定义集合时指定集合元素的类型,形如"List＜String＞ list;"的定义形式,限定列表 list 中的元素必须为 String 类型。下面的代码示范了未引入泛型之前的映射操作。

```java
Map map = new HashMap();
Student s1= new Student(1,"李白");
map.put(s1.getNo(), s1);
map.put("3","3333");
```

```
for(Object key:map.keySet())
    System.out.println(map.get(key));
```

映射的<key,value>都泛化为了 Object 类型,系统编译时对类型没有进行检查。引入泛型后,重新改写上述代码。

```
Map<Integer,Student> map = new HashMap<>();          ①
Student s1= new Student(1,"李白");
map.put(s1.getNo(), s1);
map.put("3","3333");                                  ②
for(Integer key:map.keySet())
    System.out.println(map.get(key));
```

代码行①定义并初始化键值类型分别为 Integer 和 Student 的映射 map。引入泛型后,代码行②触发编译错误,提示 put()方法的参数类型与映射 map 的键值类型不匹配。所以,泛型的引入使应用程序变得更加强壮。

9.3.8 Collections 工具类

Collections 类是 Java 提供的一个操作 Set、List 和 Map 等集合的工具类,该类中提供了大量的静态方法,可以实现排序、查询和替换等操作。

1. 排序方法

`public void sort(List<T> list)`

对列表元素按照升序进行排列,要求列表元素必须实现 java.lang.Comparable<T>接口。

`public void sort(List<T> list, Comparator<T> c)`

对列表元素按照比较器 c 定义的顺序升序排列。

(1) java.lang.Comparable<T>接口中定义比较方法 public int compareTo(T o)。比较当前对象与对象 o 之间的大小。如果当前对象大于 o,则返回正整数,等于则返回 0,否则返回负整数。

(2) Java.util.Comparator<T>接口中定义比较方法:public int compare(T o1, T o2);。比较对象 o1 与对象 o2 之间的大小。如果对象 o1 大于 o2,则返回正整数,等于则返回 0,否则返回负整数。

接口 Comparator<T>和接口 Comparable<T>都可以用于定义两个对象之间的大小关系,但是两者之间还是存在区别的:实体类实现了 Comparable<T>接口,在其内部定义排序规则,那么其自身就具有了可进行自然排序的能力,所以也称实现了 Comparable<T>接口的类拥有内部比较器;而 Comparator<T>则是在外部制定排序规则,然后作为排序策略参数传递给排序方法,如 Collections.sort(List<T> list,Comparator<T>c)。所以 Comparator<T>是一个外在的比较器,可以作为一个工具更改已经排序好的数据结构中的数据顺序,也可以赋予不具有排序功能的对象可排序的能力。

假设定义学生类 Student 具有学生编号、姓名两个属性。通常情况下,学生按照编号的升序排列,但是某些场景要求按照学生姓名的拼音升序排列。那么应该如何实现完全不同的两种排序方式呢？这时就需要分别定义内部比较器与外部比较器。例 9-6 示范了接口 Comparator<T>和 Comparable<T>的应用。

例 9-6

```
public class Student implements Comparable< Student> {
    private int no;
    private String name;
    //省略 getter、setter 及构造方法
    public String toString() {
        return name;
    }
    @Override
    public int compareTo(Student o) {
        return getNo() - o.getNo();
    }
}
```

上述代码定义了学生类 Student,实现了 Comparable<Student>接口,定义了内部排序规则,按照学生编号的升序排列。

```
public class ComparableAndComparator {
    public static void main(String[] args) {
        List<Student> stuList = new ArrayList<>();
        stuList.add(new Stu(3,"苏轼"));
        stuList.add(new Stu(1,"李白"));
        stuList.add(new Stu(2,"杜甫"));
        Collections.sort(stuList);          //按照内部排序规则对列表排序         ①
        System.out.println(stuList);
        Comparator<Student> c = new Comparator<Student>(){//定义外部比较器    ②
            @Override
            public int compare(Student o1, Student o2) {
                Collator instance = Collator.getInstance(java.util.Locale.CHINA);
                return instance.compare(o1.getName(),o2.getName());
            }
        };
        Collections.sort(stuList, c);    //按照 c 定义的外部排序策略,对列表排序 ③
        System.out.println(stuList);
    }
}
```

例 9-6 的代码行①调用了 Collections.sort(stuList)方法,利用 Student 类内部定义的

比较器按照学生编号的升序排列。接下来，代码行②通过匿名类的方式定义了外部比较器对象 c，在匿名类中重写 compare() 方法。引入 Collator 类获取"中国"语言环境对应的 Collator 对象，按照姓名的拼音比较 Student 对象的大小。代码行③以外部比较器 c 为排序策略调用 sort() 方法对列表 stuList 进行了排序。

2. 查询方法

```
public int binarySearch(List<T> list, T key)
```

采用二分查找法在列表 list 中搜索元素 key，如果查询成功，则返回其索引值，否则返回负整数。要求列表元素升序排列。

```
public T max(Collection<T> coll)
```

返回集合 coll 中的最大元素，要求列表元素必须实现 Comparable 接口。

```
public T max(Collection<T> coll, Comparator<T> c)
```

返回集合 coll 中的最大元素，集合元素的大小关系由比较器 c 定义。

```
public int frequency(Collection<T> coll, Object o)
```

查询集合 coll 中元素 o 出现的次数。

```
public int indexOfSubList(List<T> source,List<T> target)
```

查找列表 source 中第一次出现列表 target 的索引。如果列表 source 不包含 target，则返回 −1。

3. 替换及更新方法

```
public boolean replaceAll(List<T> list, T oldVal, T newVal)
```

将列表 list 中所有的 oldVal 元素替换为 newVal。如果列表 list 不包含 oldVal 元素，则返回 false。

```
public boolean addAll(Collection<T> coll, T... elements)
```

将若干个元素添加到集合 coll。

```
public void fill(List<T> list, T obj)
```

使用元素 obj 填充列表 list。

```
public void copy(List<T> dest, List<T> src)
```

将列表 src 中的元素复制到列表 dest。dest 列表的长度必须大于或等于 src 列表的长度，否则会触发 IndexOutOfBoundsException 异常。下面的代码示范了 copy() 方法的用法。

```
List<String> list1 = new ArrayList< >();
```

```
list1.add("aaa");
list1.add("bbb");
list1.add("ccc");        //list1:[aaa, bbb, ccc]
List<String> list2 = new ArrayList< > ();
list2.add("111");
list2.add("222");
list2.add("333");
list2.add("444");        //list2:[111, 222, 333, 444]
Collections.copy(list2, list1);    //list2:[aaa, bbb, ccc, 444]
```

9.4 个人通讯录(四)设计步骤

在个人通讯录(三)的基础上使用列表 List 替代数组,实现联系人及联系人电话号码的存储。

9.4.1

9.4.1 重构联系人 Contract 类

(1) Contract 实现 Comaprable 接口。
完成联系人对象按照姓名的拼音进行比较。
(2) 基于 List 接口重构联系人 Contract 中的电话号码属性。

```java
public class Contract implements Comparable<Contract>{
    private List<String> phones;
    public Contract(String name, String gender, String email
                    , List<String> phones) throws Exception {
        setName(name);
        setGender(gender);
        setEmail(email);
        setPhones(phones);
    }
    public int compareTo(Contract o) {
        Collator instance = Collator.getInstance(java.util.Locale.CHINA);
        return instance.compare(this.getName(),o.getName());
    }
    public Contract(String name, List<String> phones) throws Exception {
        this(name, "", "", phones);
    }
    ...
}
```

(3) 重写 Contract 类的 equals()及 hashCode()方法,实现联系人对象的比较。
重写 Contract 类的 equals()及 hashCode()方法,要保证参与比较的两个对象之间的 equals()方法与 hashCode()方法的返回值是一致的。equals()方法中使用 obj instanceof

Contract，以确保参与比较的对象 obj 必须是 Contract 类或者其子类对象。

```java
public boolean equals(Object obj) {
    if(this == obj)
        return true;
    if(obj != null && obj instanceof Contract)    {
        Contract c = (Contract)obj;
        return this.getName().equals(c.getName());
    }
    return false;
}
public int hashCode() {
    return this.getName().hashCode();
}
```

（4）重写合并联系人的 mergeContract(Contract o) 方法。

假设当前联系人的电话号码列表为 p1，联系人 o 的电话号码列表为 p2，合并操作实质为获取 p1 与 p2 的并集 p1∪p2。首先利用 p1.removeAll(p2) 方法得到集合 p1－p1∩p2，然后使用 p1.addAll(p2) 方法，得到 p1∪p2。

```java
public void mergeContract(Contract o) {
    if(getName().equals(o.getName())) {
        if(getGender().equals(""))
            try {
                setGender(o.getGender());
            } catch (GenderException e) {
                e.printStackTrace();
            }
        if(this.getEmail().equals(""))
            try {
                setEmail(o.getEmail());
            } catch (EmailException e) {
                e.printStackTrace();
            }
        List<String> srcPhones = getPhones();
        srcPhones.removeAll(o.getPhones());
        srcPhones.addAll(o.getPhones());
    }
}
```

（5）重构修改联系人的 update() 方法。

```java
public boolean update(Contract c) throws Exception{
    if (getName().equals(c.getName())) {
        if (c.getEmail() != null && c.getEmail() != "")
            setEmail(c.getEmail());
```

```
            if (c.getGender() != null && c.getGender() != "")
                setGender(c.getGender());
            if (c.getPhones() != null && c.getPhones().size() != 0)
                setPhones(c.getPhones());
            return true;
        } else
            return false;
    }
```

9.4.2 重构 Family 类

重新定义 Family 类的构造方法，其他方法不需要修改。

```
public Family(String name, List<String> phones) throws Exception {
    super(name, phones);
}
public Family(String name, String gender, String email,
        List<String> phones, Date birthday, String address) throws Exception {
    super(name, gender, email, phones);
    setBirthday(birthday);
    setAddress(address);
}
```

9.4.3 重构 Partner 类

重新定义 Partner 类的构造方法，其他方法不需要修改。

```
public Partner(String name, List<String> phones) throws Exception {
    super(name, phones);
}
public Partner(String name, String gender, String email,
    List<String> phones, String title, Company company) throws Exception {
    super(name, gender, email, phones);
    setCompany(company);
    setTitle(title);
}
```

9.4.4 重构通讯录 PhoneBook 类

(1) 利用 List 接口定义联系人集合。
(2) 重构 setContracts 方法。
使用 Collections 类提供的静态方法 sort 对联系人列表进行排序。

```java
public class PhoneBook {
    private List<Contract> contracts;
    public void setContracts(List<Contract> contracts) {
        Collections.sort(contracts);
        this.contracts = contracts;
    }
    ...
}
```

（3）重构添加联系人的方法 add(Contract c)。

进行添加操作时，首先利用 Collections.binarySearch 方法在联系人列表中查找并返回联系人 c 在列表中的索引 index。如果联系人集合中不包含同名的联系人，则利用 List 的 add(-index－1, c)将联系人 c 插入有序列表；如果联系人集合中已经包含同名的联系人，则需要对联系人进行合并 mergeContract()操作。

```java
public void add(Contract c) {
    if (contracts == null) {
        contracts = new ArrayList<Contract> ();
        contracts.add(c);
        return;
    }
    int index = Collections.binarySearch(getContracts(), c);
    if(index <0){
        contracts.add(- index- 1, c);
        return;
    }else
        contracts.get(index).mergeContract(c);
}
```

（4）重构删除联系人的方法 delete(Contract c)。

```java
public boolean delete(Contract c) {
    return contracts.remove(c);
}
```

（5）重构联系人的模糊查询方法 findContractsByName(String name)。

```java
public List<Contract> findContractsByName(String name) {
    List<Contract> result = new ArrayList<Contract> ();
    for(Contract c : contracts)
        if(c.getName().contains(name))
            result.add(c);
    return result;
}
```

(6) 重构显示通讯录方法 display()。

```
public void display() {
    for (Iterator<Contract> iter = contracts.iterator(); iter.hasNext();) {
        Contract c = iter.next();
        c.display();
    }
}
```

(7) 重构修改联系人的方法 updateContract(Contract c)。

```
public boolean updateContract(Contract c) throws Exception {
    for (Contract obj:contracts) {
        if (obj.equals(c)) {
            obj.update(c);
            return true;
        }
    }
    return false;
}
```

9.5 练一练

1. 使用集合框架实现约瑟夫问题。

约瑟夫问题是指编号为 1~N 的 N 个人围坐在一起形成一个圆圈,从第 1 个人开始依次按照顺时针的方向报数,数到第 M 个人出列,输出最后剩下的一个人。

2. 编程实现惠民券的发放。

例如,某市多轮次向居民发放如表 9-1 所示的惠民券。

表 9-1 惠民券的定义及发放规则

惠民券	投放数量	初始发放比率	下一轮发放比率	说　　明
买 50 减 10	20000	2	2	
买 200 减 50	15000	2	2	
买 300 减 100	10000	1	1	
谢谢惠顾	无限	5	初始比率+该居民中奖次数	如在前 5 轮惠民券的发放过程中居民王某共中奖 2 次,则第 6 轮摇奖时,其未中奖的比率增加为 7

提示：如何根据发放比率进行抽奖？

首先,根据惠民券的发放比率计算每种惠民券(含未中奖情况)的中奖概率:"买 50 减 10"的中奖概率为 20%;"买 200 减 50"的中奖概率为 20%;"买 300 减 100"的中奖概率为 10%;"谢谢惠顾"的中奖概率为 50%;

其次，生成0~1之间的随机数；

最后计算居民获得的惠民券。

例如，生成随机数0.376，则获得惠民券"买200减50"。实现原理如下：将[0,1]分割成一个个小片段，小片段的产生与每种惠民券（含未中奖情况）的中奖概率有关，本例的实现原理图如下图所示。

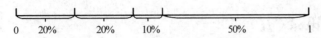

生成的随机数是0.376，落在第2个小片段上，所以该居民获得"买200减50"的惠民券。

第 10 章 基于 Swing 的图形用户界面
——个人通讯录(五)

10.1 个人通讯录(五)

设计与实现基于 GUI 的个人通讯录系统，要求：
(1) 设计与实现个人通讯录系统的图形用户界面；
(2) 利用个人通讯录(四)中实现的 PhoneBook 类实现通讯录中联系人的增、删、改、查操作。

10.2 程序设计思路

利用 Swing 组件实现通讯录系统的界面设计。通讯录系统主要定义如下用户界面原型。

1. 系统主界面

系统主界面如图 10-1 所示。

图 10-1 个人通讯录(五)主界面

主界面显示联系人姓名，并为用户提供搜索功能。用户输入待检索关键字后按 Enter 键，系统启动查询操作，将查询结果显示在下方的联系人列表中。

主界面还提供"删除联系人"及"添加联系人"功能。用户首先选中待删除的联系人，然后单击"－"按钮，待用户确认删除操作后执行删除用户操作。

用户单击"＋"按钮或者双击选中的联系人，系统进入编辑/添加联系人界面。

2. 编辑(添加)联系人界面

编辑(添加)联系人界面如图 10-2 所示。

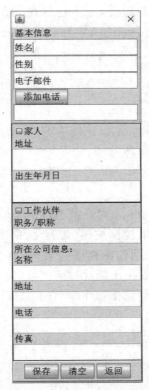

图 10-2 编辑（添加）联系人界面

10.3 关键技术

10.3.1 图形用户界面与 Swing 概述

图形用户界面(GUI)是指采用图形方式显示的用户界面,包括窗口、菜单、按钮等各种屏幕组件。用户通过键盘、鼠标与系统进行交互,与传统的命令行用户界面相比,GUI 更容易为用户所接受。

Java 语言提供了两种图形用户界面库:抽象窗口工具包(Abstract Window Toolkit, AWT)和 Swing。AWT 需要调用本地操作系统提供的方法实现功能,具有较强的平台相关性,采用 C/C++ 语言实现,属于一种重量级控件;而 Swing 是在 AWT 基础上建立的一套图形用户界面库,相较 AWT 提供更多的组件和更加完善的功能,纯 Java 实现,具有良好的平台无关性,属于轻量级控件。利用 Swing 提供的丰富、灵活的功能及模块化组件,程序员仅用很少量的代码就可以创建一个具有良好用户体验的程序界面。

图形用户界面的设计与实现通常包括选取组件、设计布局、定义事件响应这三个环节。选取组件,确定图形用户界面的基本成分;定义布局管理器,确定组件的大小及位置;最后通过事件响应定义人机交互方式。

Swing 提供了丰富的图形用户组件,大多包含在 javax.swing 包中。Java GUI 编程中的核心基类是 java.awt.Component,AWT 和 Swing 中的组件都是 Component 的子类。

10.3.2 容器

容器类 Container 也是组件 Componet 的子类,它是一种特殊组件,用来容纳组件及其他容器。Swing 主要提供以下两种容器。

1. 顶层容器

顶层容器包括主程序框架 JFrame、对话框 JDialog 及 JApplet。顶层容器相比其他组件可以独立存在,可定义标题,拥有控制按钮(最大化、最小化及关闭按钮),用户可以通过鼠标拖曳顶层容器。JFrame 主要用作主程序框架;创建 Java 小程序时使用 JApplet,用来将其嵌入浏览器窗口中;JDialog 用于创建对话框窗口。

例 10-1 示范了使用 JFrame 创建主程序窗口的方法。

例 10-1

```java
public class MainGUI extends JFrame {
    public MainGUI() {
        super("主程序窗体");                                    //定义窗体标题
        JLabel label = new JLabel("谢谢使用***应用系统");        //添加标签组件
        Container c = getContentPane();                          ①
        c.add(label);
        setBounds(100, 200, 500, 300);                           //设置窗口的边界
        setVisible(true);                                        //定义窗口可见
        setDefaultCloseOperation(JFrame.EXIT_ON_CLOSE);          //关闭窗口结束应用程序
    }
    public static void main(String[] args) {
        new MainGUI();
    }
}
```

例 10-1 的运行结果如图 10-3 所示。

图 10-3 例 10-1 的运行结果

主程序框架 JFrame 默认使用边布局管理器 BorderLayout(关于布局管理器的详细内容请参见 10.3.4 节),其初始状态不可见,所以在对主程序框架初始化时需要使用 setVisible(true) 设置窗体可见,通常需要通过 setDefaultCloseOperation() 设定主程序框架的关闭行为。顶层容器内部包含内容窗格,如例 10-1 中代码行①的容器对象 c,顶层容器中除菜单、工具栏之外的组件都需要放置在内容

窗格中。

对话框 JDialog 与 JFrame 类似,它也是一个顶层容器,相较 JFrame 更简单,无最小化、最大化按钮;可以创建具有父窗口-子窗口依赖关系的对话框;JDialog 可以分为模态对话框与非模态对话框。例 10-2 示范了使用 JDialog 创建对话框窗口的方法。

例 10-2

```
public class JDialogTest extends JFrame{
    public JDialogTest() {
        super("对话框测试");
        JDialog dialog = new JDialog(JDialogTest.this,"通知",false);        ①
        dialog.getContentPane().add(new JLabel("通知详情"));
        dialog.setBounds(50, 50, 100, 150);
        dialog.setVisible(true);
        getContentPane().add(new JLabel("欢迎进入****系统"));
        setBounds(50, 50, 400, 300);
        setVisible(true);
        setDefaultCloseOperation(JFrame.EXIT_ON_CLOSE);
    }
    public static void main(String[] args) {
        new JDialogTest();
    }
}
```

例 10-2 的运行结果如图 10-4 所示。

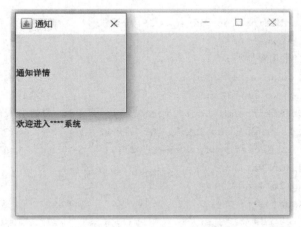

图 10-4　例 10-2 的运行结果图

例 10-2 代码行①调用了 JDialog 构造方法:

```
public JDialog(Frame owner,String title,boolean modal);
```

其中,参数 owner 对应当前对话框的父窗体,title 定义对话框的标题,当参数 modal＝true 时,创建模态对话框,否则定义非模态对话框。

2. 中间容器

中间容器主要包括 JPanel、JScrollPane、JTabbedPane、Box 等。中间容器不能独立存在，需要放置在其他容器组件中。

(1) 面板 JPanel。

面板没有标题，不可以添加菜单组件，默认使用流式布局 FlowLayout。JPanel 主要用于将界面分组，以便于分别设置布局方式。

(2) 滚动面板 JScrollPane。

相较 JPanel，滚动面板 JScrollPane 仅可以容纳一个组件，并提供水平和垂直滚动条。通常用于在有限的空间中展示数据内容较多的组件，如文本区域（JTextArea）、表格（JTable）等。

(3) 选项卡面板 JTabbedPane。

选项卡面板包含若干选项卡组件，允许用户通过单击标题或者图标的方式在一组选项卡组件之间进行切换显示。

JTabbedPane 类的常用构造方法为

```
public JTabbedPane(int tabPlacement)
```

参数 tabPlacement 定义选项卡面板放置的位置，取值为 JTabbedPane.TOP、JTabbedPane.BOTTOM、JTabbedPane.LEFT 或者 JTabbedPane.RIGHT。

例 10-3 示范了上述中间容器的使用方法。

例 10-3

```
public class ContrainerTest extends JFrame {
    public ContrainerTest() {
        super("常用容器");
        JPanel page1 = new JPanel();              //第一选项卡面板
        JPanel pal = new JPanel();                //容纳姓名、性别标签的面板
        pal.setBackground(Color.lightGray);       //设置背景色
        JLabel lblName = new JLabel("汪东颖");
        JLabel lblGender = new JLabel("女");
        pal.add(lblName);
        pal.add(lblGender);
        page1.add(pal);
        JTextArea resume = new JTextArea(5,20);   //定义文本区域
        resume.setText("工作经历\n2008.7至今 中国工商银行\n学习经历\n2000- 2008 中央财经大学\n自我评价\n");
        page1.add(new JScrollPane(resume));
        JPanel page2 = new JPanel();              //第二选项卡面板
        JTabbedPane main = new JTabbedPane();     //选项卡容器
        main.add("基本信息",page1);
        main.add("获奖信息", page2);
```

```
        Container c = this.getContentPane();
        c.add(main);
        this.setBounds(100, 200, 300, 250);
        this.setVisible(true);
        this.setDefaultCloseOperation(JFrame.EXIT_ON_CLOSE);
    }
    public static void main(String[] args) {
        new ContrainerTest();
    }
}
```

例 10-3 的运行结果如图 10-5 所示。

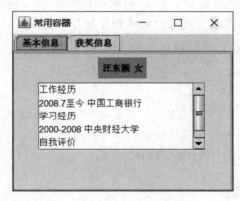

图 10-5　例 10-3 的运行结果

3. 盒子面板 Box

盒子面板与其他容器不同,它只能采用 BoxLayout 布局管理器。BoxLayout 是指容器可以沿水平或垂直方向排列多个组件。

Box 类提供以下静态方法,可以分别定义水平盒子和垂直盒子。

```
public static Box createHorizontalBox()
public static Box createVerticalBox()
```

Box 中可以定义 3 种不可见组件,分别为 Glue、Strut 和 RigidArea,它们被放置在组件之间,作为组件之间的间隔。

以下代码示范了利用 Glue 对象分隔组件的方法。

```
Box h = Box.createHorizontalBox();              //创建水平盒子
JButton b1 = new JButton("按钮 1");
JButton b2 = new JButton("按钮 2");
JButton b3 = new JButton("按钮 3");
h.add(b1);
h.add(Box.createHorizontalGlue());              //添加水平 Glue 对象
```

```
            h.add(b2);
            h.add(Box.createHorizontalStrut (20));      //添加宽度为 20 的水平 Strut 对象
            h.add(b3);
```

上述代码的运行结果如图 10-6 所示。

图 10-6　使用 Glue 对象的结果

"按钮 1"与"按钮 2"之间添加了 Glue 对象,"按钮 2"与"按钮 3"之间添加了宽度为 20 的 Strut 对象。当用户通过拖曳方式改变主程序窗口的宽度时,"按钮 1"与"按钮 2"之间的间距随之改变,而"按钮 2"与"按钮 3"之间的宽度保持不变。

10.3.3　常用组件

1. 标签

JLabel 对象,常用于显示文本或者图像。

2. 单行文本和密码框

单行文本 JTextField 对象和密码组件 JPasswordField 对象,用于实现单行文本或者密码信息的输入。

3. 文本区域

JTextArea 对象,实现多行文本的编辑。

4. 组合框

JCombox 对象,又称下拉列表框,用户可从下拉式列表框中选择已有的列表项。

5. 列表框

JList 对象,包括一系列的列表项,用户可从中选择一个或多个列表项。

6. 按钮

JButton 对象,用户单击按钮,系统响应按钮事件。

7. 单选按钮和复选按钮

分别对应 JRadioButton 和 JCheckBox 对象,按钮提供选中或者取消选中两种状态,用户通过单击实现状态的切换。对于单选按钮,要求将相关单选按钮放置在同一个

ButtonGroup 组中,以便实现组内单选;复选按钮则可以同时有多个选项被选中。

8. 滚动条

JSlider 对象,以图形方式展示或者设定数值,可以分别定义水平和垂直方向的滚动条。

9. 边框组件

JComponent 类提供 public void setBorder(Border border)方法,为组件设置边框(Border)。Swing 提供 BorderFactory 静态工厂类,该类定义了大量的静态工厂方法,用于生成 Border 实例。

例 10-4 示范了常用组件的使用方法。

例 10-4

```java
public class ComponentTest extends JFrame {
    private JTextField name;
    private JRadioButton male;
    private JRadioButton female;
    private JCheckBox ai;
    private JCheckBox net;
    private JCheckBox software;
    private static String[] strEdu =
        new String[] {"---最高学历---","高中","专科","本科","硕士及以上"};
    private JComboBox< String>  edu;
    private JTextArea resume;
    private JButton btnSave;
    private JButton btnCancle;
    public ComponentTest() {
        super("常用组件");
        Container c = this.getContentPane();
        c.setLayout(new FlowLayout());
        //定义图像对象,图像文件 1.png 存放在 ComponentTest.class 文件所在目录
        ImageIcon img = new ImageIcon(this.getClass().getResource("./1.png"));
        JLabel lblImg = new JLabel(img);          //图像标签
        name = new JTextField(10);
        male = new JRadioButton("男",true);
        female = new JRadioButton("女");
        ButtonGroup g = new ButtonGroup();
        g.add(male);
        g.add(female);
        JPanel dir = new JPanel();
        dir.setBorder(BorderFactory.createTitledBorder("研究方向"));
        ai = new JCheckBox("人工智能");
        net = new JCheckBox("网络与信息安全");
```

```
            software = new JCheckBox("软件工程");
            dir.add(ai);
            dir.add(net);
            dir.add(software);
            edu = new JComboBox<>(strEdu);
            resume = new JTextArea(5,20);
            resume.setText("个人简历");
            btnSave = new JButton("保存");
            btnCancle = new JButton("取消");
            c.add(lblImg);
            c.add(new JLabel("姓名"));
            c.add(name);
            c.add(new JLabel("性别"));
            c.add(male);
            c.add(female);
            c.add(dir);
            c.add(edu);
            c.add(new JScrollPane(resume));
            c.add(btnSave);
            c.add(btnCancle);
            this.setDefaultCloseOperation(JFrame.EXIT_ON_CLOSE);
            this.setBounds(100, 200, 330, 300);
            this.setVisible(true);
        }
        public static void main(String[] args) {
            new ComponentTest();
        }
    }
```

例 10-4 的运行结果如图 10-7 所示。

图 10-7　例 10-4 的运行结果

10. 对话框 JOptionPane

JOptionPane 类用于创建简单且样式固定的对话框。JOptionPane 类提供了以下静态方法，用来创建对话框。

```
public static int showConfirmDialog(Component parentComponent,
                Object message, String title, int optionType)
```

显示确认对话框，得到如下 JOptionPane 类中的常量作为返回值：YES_OPTION、NO_OPTION、CANCLE_ OPTION、OK_ OPTION 和 CLOSED_ OPTION。

```
public static String showInputDialog(Component parentComponent,
                Object message, Object initialSelectionValue)
```

显示输入对话框。该方法返回用户的输入值。

```
public static void showMessageDialog(Component parentComponent,
                Object message, String title, int messageType)
```

显示消息对话框。

方法 show***Dialog() 的参数 parentComponent 指定对话框的父组件；参数 message 指定对话框中显示的消息；参数 title 定义对话框的标题；参数 optionType 设置对话框的按钮类型，可以取值为 JOptionPane 类中的常量：YES_ NO_ OPTION、YES_ NO_ CANCEL_OPTION 或者 OK_CANCEL_OPTION 之一；参数 initialSelectionValue 定义输入对话框的初始值；参数 messageType 设置对话框的消息类型，可以取值为 JOptionPane 类中的常量：ERROR_MESSAGE、INFORMATION_MESSAGE、WARNING_MESSAGE、QUESTION_MESSAGE 或者 PLAIN_MESSAGE 之一。

下面的代码示范了 JOptionPane 类的应用。

```
JOptionPane.showConfirmDialog(null, "确认删除文件吗?", "确认",
                JOptionPane.YES_NO_OPTION);
```

运行显示如图 10-8 所示的确认对话框。

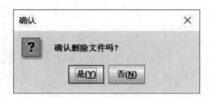

图 10-8　确认对话框

10.3.4　布局管理器

为了实现跨平台的特性，Java 使用布局管理器管理容器内所有组件的布局，如排列组件顺序、组件大小及位置等。

常用的布局管理器包括 BorderLayout、FlowLayout、GridLayout、CardLayout。布局管理器由 AWT 提供,位于 java.awt 包中。容器组件可以使用方法:

```
public void setLayout(LayoutManager manager)
```

设置容器的布局方式。

1. 边布局管理器 BorderLayout

容器被分成 5 个区域:NORTH、SOUTH、EAST、WEST 及 CENTER,如图 10-9 所示。

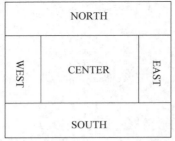

加入容器的组件需要明确指明被放置的区域,默认添加到 CENTER 区域。如果反复向同一容器区域添加组件,则后放入的组件会覆盖之前放入的组件。

BorderLayout 提供如下构造方法:

```
public BorderLayout(int hgap, int vgap)
```

其中,参数 hgap、vgap 分别表示各区域之间水平、垂直方向的间隔,默认情况下区域之间间隔为 0。

图 10-9　BorderLayout 布局示意图

2. 流式布局管理器 FlowLayout

容器内组件按照从上到下、从左到右的顺序放置。当前行无法容纳组件时,就从下一行开始继续排列组件。

FlowLayout 提供如下构造方法。

```
public FlowLayout(int align, int hgap, int vgap)
```

其中,参数 align 指定组件的对齐方式,可以取值为 FlowLayout 类的常量:LEFT、RIGHT、CENTER、LEADING 或者 TRAILING 之一。参数 hgap、vgap 分别表示各组件之间水平、垂直方向的间隔。

3. 网格布局管理器 GridLayout

容器中各组件按照从上到下、从左到右的顺序放置,与流式布局不同,各个组件呈网格状分布,大小相同。

GridLayout 提供如下构造方法。

```
public GridLayout(int rows, int cols, int hgap, int vgap)
```

其中,参数 rows、cols 指定容器被分割为 rows 行×cols 列个网格。参数 hgap、vgap 分别表示各组件之间水平、垂直方向的间隔。

例 10-5 示范了边界布局、网格布局的应用方法。

例 10-5

```java
public class GridLayoutTest extends JFrame{
    private static String[] KEY = new String[]
        {"0","1","2","3","4","5","6","7","8","9",".","+","-","*","/"};
    private JTextField srn = new JTextField(25);
    public GridLayoutTest() {
        super("计算器");
        Container c = this.getContentPane();
        srn.setBackground(Color.LIGHT_GRAY);
        c.add(srn,BorderLayout.NORTH);
        JPanel center = new JPanel();
        center.setLayout(new GridLayout(3,5));
        for(String k:KEY) {
            JButton btn = new JButton(k);
            center.add(btn);
        }
        c.add(center);
        JButton equal = new JButton("=");
        c.add(equal,BorderLayout.EAST);
        pack();
        this.setVisible(true);
        this.setDefaultCloseOperation(JFrame.EXIT_ON_CLOSE);
    }
    public static void main(String[] args) {
        new GridLayoutTest();
    }
}
```

例 10-5 的运行结果如图 10-10 所示。

4. 卡片布局管理器 CardLayout

容器中的组件被当作一叠卡片摞在一起，每次只能展示最上面的一张卡片。

CardLayout 提供如下构造方法。

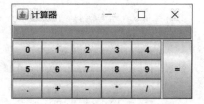

图 10-10　例 10-5 的运行结果

```java
public CardLayout(int hgap, int vgap)
```

其中，参数 hgap、vgap 分别表示各卡片之间水平、垂直方向的间隔。

CardLayout 还提供如下方法以实现卡片的显示。

```java
public void first(Container parent)      //显示容器的第一张卡片
public void next(Container parent)       //循环显示下一张卡片
public void previous(Container parent)   //循环显示上一张卡片
public void last(Container parent)       //显示容器的最后一张卡片
```

```
    public void show(Container parent,String name)    //显示容器中名字为name的卡片
```
例10-6示范了卡片布局的应用方法。

例10-6

```java
public class CardLayoutTest extends JFrame implements ActionListener{
    private JPanel center = new JPanel();
    private CardLayout card = new CardLayout();
    public CardLayoutTest() {
        super("卡片布局");
        center.setLayout(card);          //定义卡片布局
        String[] imgPath = new String[]{"../../a1.png","../../a2.png","../../
                                a3.png","../../a4.png"};
        String[] cmdStr = new String[] {"第一张","上一张","下一张","最后一张"};
        for(int i= 0;i<imgPath.length;i++) {
            ImageIcon icon = new ImageIcon(this.getClass().getResource
                                                (imgPath[i]));
            JLabel lbl = new JLabel(icon);
            center.add(""+ (i+1),lbl);   //为每个图像标签命名,并加入容器center
        }
        getContentPane().add(center);
        JPanel cmdPal = new JPanel();    //命令面板
        for(String cmd: cmdStr) {
            JButton btn = new JButton(cmd);
            btn.addActionListener(this);
            cmdPal.add(btn);
        }
        card.show(center, "3");          //显示名字为"3"的卡片
        getContentPane().add(cmdPal,BorderLayout.SOUTH);
        pack();
        setVisible(true);
        setDefaultCloseOperation(JFrame.EXIT_ON_CLOSE);
    }
    public static void main(String[] args) {
        new CardLayoutTest();
    }
    @Override
    public void actionPerformed(ActionEvent e) {
        String btnName = e.getActionCommand();
        switch(btnName) {
            case "第一张":card.first(center);break;
            case "上一张":card.previous(center);break;
            case "下一张":card.next(center);break;
            case "最后一张":card.last(center);
        }
    }
}
```

例 10-6 的运行结果如图 10-11 所示。

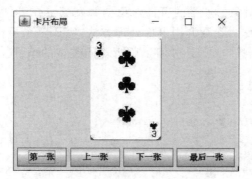

图 10-11　例 10-6 的运行结果

5. 盒子布局管理器 BoxLayout

容器中的多个组件沿水平方向或垂直方向排列。当组件的总宽度或总高度超过容器的宽度或高度时，部分组件将不可见。

BoxLayout 提供如下构造方法。

```
public BoxLayout(Container target, int axis)
```

其中，参数 target 表示容器，axis 定义排列方向，常用的取值为 BoxLayout 类的常量 X_AXIS 与 Y_AXIS。

10.3.5　事件处理

1. 事件处理流程

为了使 GUI 能够响应用户的操作，必须为组件添加事件处理机制。事件处理中主要涉及三类对象：事件源、事件及事件监听器。

事件源：产生事件的对象，通常指 GUI 中的组件，如按钮、文本框等。

事件：指在事件源上发生的事情，通常代表用户的一次操作，如单击或者输入等。

事件监听器：实现了监听器接口的特殊类，负责监听事件，并在事件发生时通过事件处理方法对事件做出响应。

图 10-12 显示了事件处理的流程示意。

图 10-12　事件处理流程示意图

当事件源上的某个触发动作发生时,系统会生成相应的事件对象,如用户单击命令按钮会生成 ActionEvent 对象。产生的事件对象会触发绑定在事件源上的事件监听器,事件监听器会以事件对象作为参数调用事件处理方法,完成整个事件处理过程。所谓事件监听器,就是实现监听器接口的类,可用于监听不同类型的事件。例 10-7 示范了事件处理的应用。

例 10-7

```java
public class MainWindow extends JFrame{                     //主程序窗体
    private static String[] province = new String[] {"- 省份- ","山东省","河北省"};
    private static String[][] city = new String[][] {
        {"--城市--","济南","青岛","淄博","枣庄","东营","烟台"},
        {"--城市--","石家庄","唐山","邯郸","秦皇岛","保定"}};
    private JComboBox<String> cmbPro;
    private JComboBox<String> cmbCity;
    private JLabel lblSelected;
    public void init() {                                     //初始化方法
        this.setTitle("ItemEvent 事件");
        cmbPro = new JComboBox<>(province);
        cmbCity = new JComboBox<>();
        lblSelected = new JLabel();
        cmbPro.addItemListener(new ProvinceList());          //绑定监听器实例    ①
        cmbCity.addItemListener(new CityList());             //绑定监听器实例    ②
        Container c = getContentPane();
        c.setLayout(new FlowLayout());
        c.add(cmbPro);
        c.add(cmbCity);
        c.add(lblSelected);
        setVisible(true);
        setSize(300,200);
        setDefaultCloseOperation(JFrame.EXIT_ON_CLOSE);
    }
    class ProvinceList implements ItemListener{      //定义内部类实现监听器接口    ③
        @Override
        public void itemStateChanged(ItemEvent e) {                              ④
            if(e.getStateChange()==ItemEvent.SELECTED) {                         ⑤
                JComboBox<String> com = (JComboBox<String>)e.getSource();
                int index = com.getSelectedIndex();
                if(index != 0) {
                    cmbCity.removeAllItems();
                    for(String c:city[index-1])
                        cmbCity.addItem(c);
                    String str= (String)com.getSelectedItem();
                    lblSelected.setText(str);
                }
```

```
            }
        }
    }
    class CityList implements ItemListener{
        @Override
        public void itemStateChanged(ItemEvent e) {
            if(e.getStateChange()==ItemEvent.SELECTED) {
                JComboBox<String> com = (JComboBox<String>)e.getSource();
                if(com.getSelectedIndex() != 0) {
                    String str=(String)com.getSelectedItem();
                    lblSelected.setText(cmbPro.getSelectedItem()+" "+str);
                }
            }
        }
    }
}
public class ItemEventTest {
    public static void main(String[] args) {
        MainWindow main = new MainWindow();
        main.init();
    }
}
```

例 10-7 的运行结果如图 10-13 所示。

例 10-7 中的代码行①为事件源（组合框 cmbPro）绑定监听器对象。监听器 ProvinceList 实现 ItemListener 接口，如代码行③所示。上述代码将监听器 ProvinceList 定义为内部类，这是因为监听器类通常都需要访问或者修改用户界面组件，定义为内部类就可以直接访问外部类（用户界面）的变量和方法。

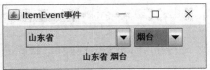

图 10-13　例 10-7 的运行结果

代码行④定义 ProvinceList 监听器的事件处理方法 itemStateChanged()。程序执行时，当用户改变组合框 cmbPro 的选择项时，系统会生成 ItemEvent 事件并将之传递给监听器对象，事件监听器则会调用事件处理方法 itemStateChanged()。

注意：当用户改变组合框 cmbPro 的选择项时，会同时触发两次 ItemEvent 事件，一个选择框从未选中状态变为选中状态；同时，另一个选择框从选中状态变为未选中状态。此处仅对选中状态的选择框进行处理，所以代码行⑤增加了判断语句。

2. 事件与事件监听器

通过上述事件处理流程的分析，可以得到如下 Java 程序事件处理的模板。

```
class XXXXListener implements XXXListener {   //定义监听器
    @Override
```

```
        public void xxx(XxxEvent e) {            //定义一个或若干事件处理方法
            ...
        }
    }
    yyy.addXXXListener(new XXXXListener());      //事件源 yyy 绑定监听器对象
```

创建监听器类 XXXXListener,该类实现监听器接口 XXXListener,重新定义事件处理方法 xxx(),使得可以处理对应事件 XxxEvent。程序中,在事件源上使用 addXXXListener()方法绑定监听器实例。

Java 提供了丰富的事件类,用于封装不同事件源上发生的特定操作;同时,系统也定义了大量的事件监听器接口,用于监听不同类型的事件。具体的事件、监听器接口及事件处理方法请参见 JavaAPI 文档。

3. 事件监听器的实现方式

(1) 内部类监听器。

例 10-7 中的事件监听器就采用了内部类的定义方式。通过内部类定义事件监听器可以方便地访问 GUI 主类中定义的属性和方法,包括私有的属性与方法;不足之处在于,过多的内部监听器类使得 GUI 主类过于庞杂,并且内部监听器不可以在不同界面中进行复用。

(2) 匿名内部类监听器。

事件监听器的定义还经常采用匿名内部类的定义方式。例如 10-7 中的代码行①可重写为如下形式。

```
cmbPro.addItemListener(new ItemListener() {
    @Override
    public void itemStateChanged(ItemEvent e) {
        if(e.getStateChange()==ItemEvent.SELECTED) {
            JComboBox<String> com = (JComboBox<String>)e.getSource();
            int index = com.getSelectedIndex();
            if(index != 0) {
                cmbCity.removeAllItems();
                for(String c:city[index-1])
                    cmbCity.addItem(c);
                String str=(String)com.getSelectedItem();
                lblSelected.setText(str);
            }
        }
    }
});
```

匿名内部类的实现方式与内部类的实现方式类似,便于在事件监听器中访问 GUI 主类的方法和属性;不足之处在于 GUI 主类太臃肿,并且内部监听器不可以在不同界面中进行复用。

(3) GUI 界面作为监听器。

GUI 界面可以直接实现监听器接口,作为事件监听器类。例 10-7 采用 GUI 界面作为监听器的定义如下。

```java
public class MainWindow extends JFrame implements ItemListener {
    //省略与程序例 10-7 相同的代码行
    public void init() {                              //初始化方法
        cmbPro.addItemListener(this);                 //绑定监听器实例
        cmbCity.addItemListener(this);                //绑定监听器实例
    }
    @Override
    public void itemStateChanged(ItemEvent e) {
        if (e.getStateChange() == ItemEvent.SELECTED) {
            JComboBox<String> com = (JComboBox<String>) e.getSource();
            if (com == cmbPro) {                      //cmbPro 组件的 ItemEvent 事件处理
                int index = com.getSelectedIndex();
                if (index != 0) {
                    cmbCity.removeAllItems();
                    for (String c : city[index - 1])
                        cmbCity.addItem(c);
                    String str = (String) com.getSelectedItem();
                    lblSelected.setText(str);
                }
            } else if (com == cmbCity) {              //cmbCity 组件的 ItemEvent 事件处理
                if (com.getSelectedIndex() != 0) {
                    String str = (String) com.getSelectedItem();
                    lblSelected.setText(cmbPro.getSelectedItem() + " " + str);
                }
            }
        }
    }
}
```

4. 事件适配器

创建事件监听器类时,需要实现相应监听器接口中的所有方法,如 WindowListener 接口内定义了 windowActivated()等 7 个方法。但是应用程序往往只对某些方法感兴趣,这时就需要在应用程序中书写很多空实现的方法。为了精简事件监听器的空实现方法,Java 引入了事件适配器。事件适配器是一种特殊的事件监听器,其实现了事件监听器接口,并为接口中的每个方法提供空实现。这样,当需要创建事件监听器时,就可以通过继承事件适配器实现。事件适配器为所有方法提供了空实现,所以应用程序只需要实现感兴趣的方法即可。

例 10-8 示范了通过事件适配器创建事件监听器的方法。

例 10-8

```java
public class WindowListenerTest{
    public static void main(String[] args) {
        JFrame main = new JFrame("WindowEvent事件");
        main.setSize(300, 100);
        main.setVisible(true);
        main.setDefaultCloseOperation(JFrame.DO_NOTHING_ON_CLOSE);
        main.addWindowListener(new WindowAdapter() {
            public void windowClosing(WindowEvent e) {
                int result = JOptionPane.showConfirmDialog(null,
                                                "确认退出应用程序吗?");
                if(JOptionPane.YES_OPTION == result)
                    System.exit(1);
                else
                    return;
            }
        });
    }
}
```

执行应用程序,用户单击窗体的关闭按钮,弹出如图 10-14 所示的对话框。

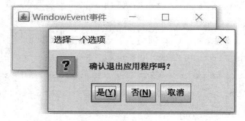

图 10-14　例 10-8 的运行结果

用户单击"是"按钮,退出应用程序,否则关闭对话框,返回主程序窗体。

例 10-8 通过继承事件适配器 WindowAdapter 创建了事件监听器,其中仅需要重写 windowClosing()方法即可。

10.3.6　菜单组件

菜单是 GUI 中常用的组件之一。菜单组件与其他常用组件不同,它不能放置在普通容器中,只能放置在固定的菜单容器(菜单栏或弹出式菜单)中,且不受布局管理器的约束。

菜单组件由菜单栏、菜单和菜单项三部分组成。图 10-15 显示了 Swing 菜单组件类之间的继承与组合关系。

如图 10-15 所示,一个菜单栏由若干个菜单组成,一个菜单又包括若干个菜单项。菜

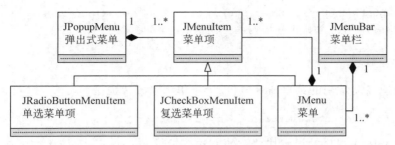

图 10-15 菜单组件类

单项(JMenuItem)是菜单(JMenuItem)、复选菜单项(JCheckBoxMenuItem)和单选菜单项(JRadioButtonMenuItem)的共同父类。菜单栏也称下拉式菜单,通过调用 JFrame 类的 setJMenuBar()方法可以将下拉式菜单放置在 JFrame 窗口中。Java Swing 中还存在一种菜单形式——弹出式菜单(JPopupMenu),弹出式菜单由若干个菜单项组合而成,可以放置在 GUI 的任何位置。

1. 菜单栏 JMenuBar

JMenuBar 类提供如下常用方法。

```
public JMenuBar()                //JMenuBar 的构造方法,用于创建菜单栏对象
public JMenu add(JMenu c)        //将菜单对象添加到菜单栏实例中
```

2. 菜单 JMenu

JMenu 类提供如下常用方法。

```
public JMenu(String s)                       //JMenu 类的构造方法,创建标题为 s 的菜单对象
public JMenuItem add(JMenuItem menuItem)     //将菜单项对象添加到菜单实例中
public JMenu add(JMenu menu)                 //将 menu 添加到当前菜单,构建嵌套式菜单
public void addSeparator()                   //将分隔线添加到菜单,实现将功能相似的菜单项分组
```

3. 菜单项 JMenuItem

JMenuItem 类提供如下常用方法。

```
public JMenuItem(String text)    //JMenuItem 类的构造方法,创建标题为 text 的菜单项
```

4. 复选菜单项 JCheckBoxMenuItem

JCheckBoxMenuItem 类作为 JMenuItem 的子类,提供如下常用方法。

```
public JCheckBoxMenuItem(String text, boolean b)
                                 //创建标题为 text、初始状态为 b 的复选菜单项对象
public boolean getState()        //返回当前复选菜单项的选择状态
```

5. 单选菜单项 JRadioButtonMenuItem

JMenuItem 的子类,JRadioButtonMenuItem 类提供如下常用方法。

```
public JRadioButtonMenuItem(String text, boolean selected)
```
//创建标题为 text、初始状态为 selected 的单选菜单项对象

6. 定义快捷键

为了快速启动常用的菜单命令,可以为菜单项设置快捷键。

```
public void setAccelerator(KeyStroke keyStroke)
```

JMenuItem 类提供的设置快捷键的方法。参数 keyStroke 对应于按下或释放某个特定的键。

菜单项的快捷键往往与 Ctrl 键、Shift 键、Alt 键等配合使用。KeyStroke 类提供下列静态方法以获取这种组合键对应的 KeyStroke 对象。

```
public static KeyStroke getKeyStroke(int keyCode, int modifiers)
```

参数 keyCode 可以取值 java.awt.event.KeyEvent 类的静态常量,如 KeyEvent.VK_A 至 KenEvent.VK_Z 分别表示按键 a~z 的键码值。参数 modifiers 表示控制键,可以取值 java.awt.event.InputEvent 类的静态常量,如 InputEvent.CTRL_DOWN_MASK 表示 Ctrl 键。

7. 下拉式菜单的创建步骤

(1) 创建菜单栏并置于 JFrame 窗体中。
JFrame 类提供如下放置菜单栏组件的方法。

```
public void setJMenuBar(JMenuBar menubar)
```

(2) 创建若干菜单并加入菜单栏。
(3) 创建若干菜单项分别加入相应的菜单对象。
(4) 为菜单项设置快捷键。
(5) 为菜单项绑定事件监听器。

例 10-9 示范了为主程序窗体添加下拉式菜单的应用。

例 10-9

```
public class JMenuTest extends JFrame implements ActionListener{
    public JMenuTest() {
        super("菜单测试");
        JMenuBar bar = new JMenuBar();        //1)创建菜单栏并置于 JFrame 窗体中
        this.setJMenuBar(bar);
        JMenu file = new JMenu("File");       //2)创建若干菜单并加入菜单栏
        JMenu edit = new JMenu("Edit");
        bar.add(file);
        bar.add(edit);
        JMenuItem fileNew = new JMenuItem("new");
                                              //3)创建菜单项并加入相应的菜单对象
```

```java
        JMenuItem fileSave = new JMenuItem("save");
        JMenuItem fileExit = new JMenuItem("exit");
        file.add(fileNew);
        file.add(fileSave);
        file.addSeparator();
        file.add(fileExit);
        JMenuItem editCopy = new JMenuItem("copy");
        editCopy.setAccelerator(KeyStroke.getKeyStroke(KeyEvent.VK_C, InputEvent.
        CTRL_DOWN_MASK));                            //4)为菜单项设置快捷键
        JMenuItem editCut = new JMenuItem("cut");
        JMenuItem editPaste = new JMenuItem("paste");
        JCheckBoxMenuItem autoSave = new JCheckBoxMenuItem("自动保存");
        ButtonGroup font = new ButtonGroup();
        JRadioButtonMenuItem font12 = new JRadioButtonMenuItem("12pt", true);
        JRadioButtonMenuItem font8 = new JRadioButtonMenuItem("8pt");
        JRadioButtonMenuItem font10 = new JRadioButtonMenuItem("10pt");
        font.add(font8);
        font.add(font10);
        font.add(font12);
        edit.add(editCopy);
        edit.add(editCut);
        edit.add(editPaste);
        edit.addSeparator();
        edit.add(autoSave);
        edit.addSeparator();
        edit.add(font8);
        edit.add(font10);
        edit.add(font12);
        fileNew.addActionListener(this);              //5)为菜单项绑定事件监听器
        fileSave.addActionListener(this);
        fileExit.addActionListener(this);
        editCopy.addActionListener(this);
        editCut.addActionListener(this);
        editPaste.addActionListener(this);
        JTextArea text = new JTextArea();
        getContentPane().add(new JScrollPane(text));
        setVisible(true);
        setSize(300, 200);
        setDefaultCloseOperation(JFrame.EXIT_ON_CLOSE);
    }
    public static void main(String[] args) {
        new JMenuTest();
    }
    @Override
```

```java
    public void actionPerformed(ActionEvent e) {
        String cmd = e.getActionCommand();
        switch(cmd) {
        case "new":System.out.println("new");break;
        case "save":System.out.println("save");break;
        case "exit":System.exit(0);
        case "copy":System.out.println("copy");break;
        case "cut":System.out.println("cut");break;
        case "paste":System.out.println("paste");break;
        }
    }
}
```

例 10-9 的运行结果如图 10-16 所示。

8. 弹出式菜单

弹出式菜单又称右键菜单，一般是通过在容器界面上单击鼠标右键而弹出的浮动菜单。弹出式菜单使用 JPopupMenu 对象表示。创建弹出式菜单的步骤如下。

（1）创建 JPopupMenu 对象。

（2）创建若干菜单项 JMenuItem 对象，依次加入 JPopupMenu 对象。

图 10-16　例 10-9 的运行结果

（3）为需要弹出式菜单的组件编写鼠标监听器，当用户释放右键时弹出 JPopupMenu 对象。

（4）为菜单项 JMenuItem 对象绑定事件监听器。

例 10-10 在例 10-9 的基础上为窗体中的 JTextArea 组件添加了弹出式菜单。

例 10-10

```java
public class JMenuTest extends JFrame implements ActionListener{
    public JMenuTest() {
    //省略与程序例 10-11 相同部分的代码
        JPopupMenu popupMenu= new JPopupMenu( );              //创建 JPopupMenu 对象
        JMenuItem editC = new JMenuItem("copy");
        JMenuItem editX = new JMenuItem("cut");
        JMenuItem editV = new JMenuItem("paste");
        popupMenu.add(editC);
        popupMenu.add(editX);
        popupMenu.add(editV);
        editC.addActionListener(this);
        editX.addActionListener(this);
        editV.addActionListener(this);
        text.addMouseListener(new MouseAdapter(){     //编写鼠标监听器,显示弹出式菜单
```

```
            public void mouseReleased( MouseEvent e ){
                    if (e.isPopupTrigger())         //若释放鼠标右键
                    popupMenu.show(e.getComponent(),e.getX(),e.getY());
            }
        }
    );
}//JMenuTest 构造方法定义结束
} //类 JMenuTest 定义结束
```

上述代码行中定义的 JMenuItem 对象将与例 10-9 的下拉菜单中定义的 JMenuItem 对象共享 ActionEvnet 事件监听器。

执行应用程序,用户在 JTextArea 组件上单击鼠标右键,弹出右键菜单,运行结果如图 10-17 所示。

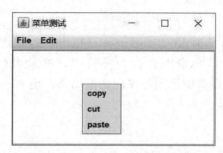

图 10-17 例 10-10 的运行结果

10.3.7 表格组件

表格是图形用户界面中的常用组件。表格通常由表头和表格数据组成,表头也称表的标题,用来定义每列的名称。表格数据定义了表格的具体内容,由若干单元格组成。表格组件不仅可以展示数据,还可以对表格数据进行编辑操作。Java 提供了 JTable 及相关类,用于构建及编辑表格组件。

1. JTable 类

JTable 类用于创建表格组件,常用的方法定义如下。

public JTable(int numRows,int numColumns)

创建 numRows 行、numColumns 列的表格,列标题默认为"A""B""C"等。

public JTable(Object[][] rowData, Object[] columnNames)

以 columnNames 为表头、以数组 rowData 为表格数据创建表格。

public JTable(Vector<? extends Vector>rowData, Vector<?>columnNames)

以一维向量 columnNames 为表头、二维向量 rowData 为表格数据创建表格。

```
public JTable(TableModel dm)
```
使用指定的数据模型 dm 创建表格。

```
public int getSelectedColumn()
public int getSelectedRow()
```
获取当前选中的第一列(或者行)的索引。如果没有列(或者行)被选中,则返回-1。

```
public int getSelectedColumnCount()
public int getSelectedRowCount()
```
获取当前选中行(或者列)数。

```
public TableModel getModel()
public void setModel(TableModel model)
```
获取或者设置表格的数据模型,其中参数 model 指定数据模型。

表格组件的列标题、单元格内容皆为字符串。通常表格对象被放置在滚动面板 JScrollPane 中,为表格数据提供横向与纵向的滚动条。例 10-11 使用 JTable 类创建了一个简单的二维表格。

例 10-11

```
public class JTableTest extends JFrame {
    public JTableTest() {
        super("简单的表格");
        String[] columnNames = new String[] {"学号","姓名","性别"};
        String[][] rowData = new String[][] {{"20160101","刘鑫","男"},
            {"201601021","王伟长","男"},{"201601031","孙玮","女"}};
        JTable table = new JTable(rowData,columnNames);
        getContentPane().add(new JScrollPane(table));
        pack();
        setDefaultCloseOperation(JFrame.EXIT_ON_CLOSE);
        setVisible(true);
    }
    public static void main(String[] args) {
        new JTableTest();
    }
}
```

2. TableModel 接口

Swing 组件采用 MVC(Model View Controller,模型-视图-控制器)设计模式,将 GUI 组件的显示逻辑和数据逻辑进行分离,对于相同的数据,允许程序员自定义不同的显示逻辑以改变 GUI 组件的显示外观,提供更多的灵活性。相对于 JTable 组件,TableModel 实现了表格组件的数据逻辑。TableModel 接口封装了表格中的各种数据及对数据进行操作的方法。抽象类 AbstractTableModel 是 TableModel 接口的一个实现类,其为 TableModel 接口中的大多数方法提供默认实现。应用程序可以利用抽象类 AbstractTableModel 的子类 DefaultTableModel 直接操作表格组件。类 DefaultTableModel 提供如下常用方法。

```
public DefaultTableModel(int numRows,int numColumns)
public DefaultTableModel(Object[][] rowData, Object[] columnNames)
public DefaultTableModel(Vector<?extends Vector>rowData, Vector<?>columnNames)
```

DefaultTableModel 常用的构造方法。参数的定义与 JTable 类构造方法中参数的定义相同。

```
public void addColumn(Object columnName)
public void addColumn(Object columnName,Vector<?>columnData)
public void addColumn(Object columnName,Object[] columnData)
```

将列 columnName 作为最后一列添加至表格。若没有指定参数 columnData,则新添加的数据列为空,否则定义 columnName 列的数据为数组(或者向量)columnData。

```
public void addRow(Vector<?>rowData)
public void addRow(Object[] rowData)
```

将行 rowData 作为最后一行添加至表格。

```
public void removeRow(int row)
```

删除表格的 row 行。DefaultTableModel 类没有提供删除表格列的方法。

```
public Object getValueAt(int row,int column)
public void setValueAt(Object aValue,int row,int column)
```

获取或者设置 row 行、column 列单元格数据。参数 aValue 指定设置单元格的内容。

```
public int getColumnCount()
public void setColumnCount(int columnCount)
```

获取或者设置表格的列数。若参数 columnCount 大于表格的现有列数,则在表格的末尾添加空的数据列;若参数 columnCount 小于表格的现有列数,则舍弃多余的数据列。

```
public int getRowCount ()
public void setRowCount(int rowCount)
```

获取或者设置表格的行数。若参数 rowCount 大于表格的现有行数,则在表格的末尾添加空的数据行;若参数 rowCount 小于表格的现有行数,则舍弃多余的数据行。

```
public String getColumnName(int column)
```

获取表格 column 列的标题。

```
public Vector<Vector>getDataVector()
public void setDataVector(Vector<?extends Vector> dataVector,
         Vector<?>columnIdentifiers)
```

获取或者设置表格数据向量。参数 dataVector、columnIdentifiers 分别表示替换之后的表格数据和表头向量。下面例 10-12 示范了通过 DefaultTableModel 类提供的方法实现表格的操作。

例 10-12

```java
public class DefaultTableModelTest extends JFrame {
    public DefaultTableModelTest() {
        super("表格数据模型");
        DefaultTableModel model = new DefaultTableModel();
        model.addColumn("学号");                //添加表列
        model.addColumn("姓名");
        model.addColumn("性别");
        model.addRow(new String[] {"20160101","刘鑫","男"});     //添加表行
        model.addRow(new String[] {"20160102","王伟长","男"});
        model.addRow(new String[] {"20160103","孙玮","女"});
        JTable table = new JTable(model);       //利用 TableModel 创建表格
        JPanel palCmd = new JPanel();
        palCmd.add(new JLabel("学号"));
        JTextField tfdNo = new JTextField(10);
        palCmd.add(tfdNo);
        palCmd.add(new JLabel("姓名"));
        JTextField tfdName = new JTextField(10);
        palCmd.add(tfdName);
        palCmd.add(new JLabel("性别"));
        JTextField tfdGender = new JTextField(10);
        palCmd.add(tfdGender);
        JButton btnUpd = new JButton("更新");
        JButton btnDel = new JButton("删除");
        JButton btnNew = new JButton("添加");
        palCmd.add(btnNew);
        palCmd.add(btnDel);
        palCmd.add(btnUpd);
        getContentPane().add(new JScrollPane(table));
        getContentPane().add(palCmd,BorderLayout.NORTH);
        pack();
        setDefaultCloseOperation(JFrame.EXIT_ON_CLOSE);
        btnUpd.addActionListener(new ActionListener() {
            @Override
            public void actionPerformed(ActionEvent e) {
                int r = table.getSelectedRow();
                if(r!=-1) {    //更新选定行
                    model.setValueAt(tfdNo.getText(), r, 0);
                    model.setValueAt(tfdName.getText(), r, 1);
                    model.setValueAt(tfdGender.getText(), r, 2);
                }
```

```java
            }
        });
        btnDel.addActionListener(new ActionListener() {
            @Override
            public void actionPerformed(ActionEvent e) {
                if(table.getSelectedRow()!=-1)
                    model.removeRow(table.getSelectedRow());//删除选定行
            }
        });
        btnNew.addActionListener(new ActionListener() {
            @Override
            public void actionPerformed(ActionEvent e) {
                model.addRow(new Vector<String>());          //表格末尾添加空行
            }
        });
        table.addMouseListener(new MouseAdapter() {
                                                    //表格选中行数据填充上方单行文本
            @Override
            public void mouseClicked(MouseEvent e) {
                int r = table.getSelectedRow();
                if(r!=-1) {
                    tfdNo.setText((String)table.getValueAt(r, 0));
                    tfdName.setText((String)table.getValueAt(r, 1));
                    tfdGender.setText((String)table.getValueAt(r, 2));
                }
            }
        });
        setVisible(true);
    }
    public static void main(String[] args) {
        new DefaultTableModelTest();
    }
}
```

例 10-12 运行结果如图 10-18 所示。

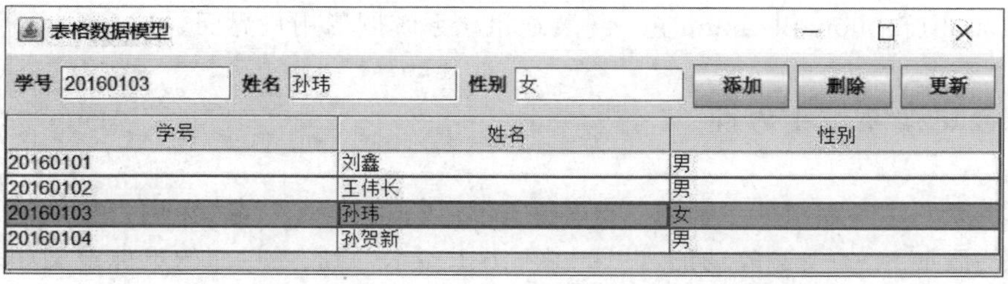

图 10-18 例 10-12 的运行结果

用户单击"添加"按钮,则在表格的末尾添加空行;用户选中表格,被选中的表格行数据会显示在屏幕上方的文本框中;用户输入或者修改文本框中的数据,单击"更新"按钮,则利用文本框数据更新表格被选中的行;用户单击"删除"按钮,则表格被选中的行被删除。

10.4 个人通讯录(五)设计步骤

10.4.1 项目总体结构

项目文件结构如图 10-19 所示。

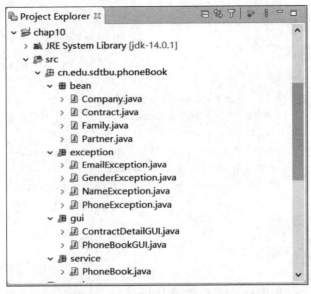

图 10-19 个人通讯录(五)项目文件结构图

cn.edu.sdtbu.phoneBook.bean 包:封装通讯录系统相关实体类。
cn.edu.sdtbu.phoneBook.exception 包:封装自定义异常类。
cn.edu.sdtbu.phoneBook.service 包:封装通讯录业务逻辑类 PhoneBook。
cn.edu.sdtbu.phoneBook.gui 包:封装通讯录系统图形用户界面。

10.4.2

10.4.2 通讯录系统主界面

通讯录系统主界面封装了通讯录业务逻辑类 PhoneBook 对象,为用户提供通讯录的显示、搜索、添加、维护及删除功能,设计如图 10-1 所示,构成主界面的组件定义如表 10-1 所示。

表 10-1　构成主界面的组件描述

组件	类型	响应事件	描述
搜索	JTextField	ActionEvent	按 Enter 键即可执行搜索功能，若搜索框为空串，则显示所有联系人
+	JButton	ActionEvent	显示"添加联系人"对话框，用户关闭"添加联系人"对话框后，将新的联系人添加到 PhoneBook 对象，并刷新界面列表
−	JButton	ActionEvent	在 PhoneBook 对象中删除当前选中的联系人，并刷新界面列表
联系人列表	JList	MouseEvent	双击联系人，显示"维护联系人信息"对话框，用户关闭"维护联系人信息"对话框后，修改 PhoneBook 对象中的联系人信息，并刷新界面列表

1. 设置 GUI 外观

通讯录系统重新设置了 Swing 组件的默认外观，通过 Javax.swing.UIManager 类提供的如下方法定制系统皮肤。

```
public static void setLookAndFeel(String className)
```

设置 GUI 的外观风格的静态方法。其中，参数 className 指定具体 LookAndFeel 子类的完全限定名称。

常见的几种外观风格如下。

默认的 Metal 风格：className="javax.swing.plaf.metal.MetalLookAndFeel"。
Windows 风格：className="com.sun.java.swing.plaf.windows.WindowsLookAndFeel"。
Nimbus 风格：className="javax.swing.plaf.nimbus.NimbusLookAndFeel"。
利用 UIManager 提供的如下静态方法为 GUI 组件设置统一的字体。

```
public static Object put(Object key,Object value)
```

例如：

```
UIManager.put("Button.font", new Font("宋体", Font.PLAIN, 15));
```

定义 GUI 所有的按钮组件具有相同的字体，即宋体、15 号字。

2. PhoneBookGUI 类的实现

PhoneBookGUI 类的详细代码如下。

```
public class PhoneBookGUI extends JFrame {
    private DefaultListModel<Contract> listModel;    //JList 组件的数据模型
    private JList<Contract> phoneList;               //JList 组件
    private PhoneBook bookServ;                      //业务逻辑对象
```

```java
private Contract currentContract;                        //当前的联系人
{//初始化块生成初始的联系人列表
    java.util.List<Contract> c = new ArrayList<Contract>();
    Set<String> list1 = new HashSet<String>();
    list1.add("13602344578");
    try {
        c.add(new Contract("王新明", list1));
    } catch (Exception e1) {
    }
    bookServ = new PhoneBook(c);                         //实例化业务逻辑对象
}                                                        //初始化块定义结束
public PhoneBookGUI() {                                  //构造方法
    super("通讯录");
    try {
        UIManager.setLookAndFeel("javax.swing.plaf.nimbus
                .NimbusLookAndFeel");                    //定制皮肤
    } catch (ClassNotFoundException | InstantiationException
            | IllegalAccessException | UnsupportedLookAndFeelException e) {
        e.printStackTrace();
    }
    //自定义各组件的字体
    UIManager.put("Button.font", new Font("宋体", Font.PLAIN, 15));
    UIManager.put("Label.font", new Font("宋体", Font.PLAIN, 15));
    JPanel top = new JPanel();
    JTextField tfdSearch = new JTextField("搜索", 20);
    JButton btnAdd = new JButton("+");
    JButton btnDel = new JButton("-");
    top.add(tfdSearch);
    top.add(btnAdd);
    top.add(btnDel);
    listModel = new DefaultListModel<Contract>();
    freshListModel(bookServ.getContracts());
    phoneList = new JList<Contract>(listModel);
    Container c = this.getContentPane();
    c.add(top, BorderLayout.NORTH);
    c.add(new JScrollPane(phoneList));
    pack();
    setVisible(true);
    setDefaultCloseOperation(JFrame.EXIT_ON_CLOSE);
    tfdSearch.addActionListener(new ActionListener() {
        @Override
        public void actionPerformed(ActionEvent e) {
            if (!tfdSearch.getText().equals("")) {
                java.util.List<Contract> founded =
                    bookServ.findContainsByName(tfdSearch.getText());
```

```java
                freshListModel(founded);
            } else
                freshListModel(bookServ.getContracts());
        }
    });
    btnAdd.addActionListener(new ActionListener() {
        @Override
        public void actionPerformed(ActionEvent e) {
            Contract c = new Contract();
            setCurrentContract(c);
            new ContractDetailGUI(PhoneBookGUI.this);
            if(c.getName()!=null && c.getName()!="")
            bookServ.addContract(getCurrentContract());
            freshListModel(bookServ.getContracts());
        }
    });
    btnDel.addActionListener(new ActionListener() {
        @Override
        public void actionPerformed(ActionEvent e) {
            Contract delContract = phoneList.getSelectedValue();
            if (null == delContract) {
                JOptionPane.showMessageDialog(PhoneBookGUI.this,
                                "请选中待删除的联系人");
                return;
            }
            if (JOptionPane.YES_OPTION = =
                    JOptionPane.showConfirmDialog(PhoneBookGUI.this,
                            "确认删除联系人吗?")) {
                bookServ.deleteContract(delContract);
                freshListModel(bookServ.getContracts());
            }
        }
    });
    phoneList.addMouseListener(new MouseAdapter() {
        @Override
        public void mouseClicked(MouseEvent arg0) {
            if (arg0.getClickCount() == 2) {
                Contract c = phoneList.getSelectedValue();
                setCurrentContract(c);
                new ContractDetailGUI(PhoneBookGUI.this);
                bookServ.updateContract(getCurrentContract());
                freshListModel(bookServ.getContracts());
            }
        }
```

```
        });
    }//构造方法结束
    private void freshListModel(java.util.List<Contract> contracts) {
        listModel.removeAllElements();
        for (Contract p : contracts) {
            listModel.addElement(p);
        }
    }
    public static void main(String[] args) {
        new PhoneBookGUI();
    }
}//PhoneBookGUI 类定义结束
```

10.4.3 "编辑(添加)联系人"对话框

在主界面双击选中的联系人或者单击"＋"按钮会弹出如图 10-2 所示的"编辑(添加)联系人"对话框。

"编辑(添加)联系人"对话框封装联系人 Contract 类的对象,实现该对象的显示及维护功能。构成联系人对话框的主要组件如表 10-2 所示。

表 10-2 构成联系人对话框的组件描述

组　件	类　型	响应事件	描　述
姓名、性别、电子邮件、地址(家庭)、出生年月日、职务/职称、公司名称、地址、电话及传真	JTextField		接收用户输入的联系人信息
联系人电话号码	JTextField	MouseEvent	单击鼠标右键,显示弹出式菜单
弹出式菜单	JPopupMenu	ActionEvent	在弹出式菜单中定义"删除电话号码"菜单项,单击此菜单项即可删除当前电话号码
添加电话	JButton	ActionEvent	添加新的单行文本组件
家人	JCheckBox	ChangeEvent	勾选"家人"复选框则取消勾选"工作伙伴"复选框,同时设置"地址""出生年月日"文本框可用;若取消勾选"家人"复选框,则将"地址""出生年月日"文本框清空并设置为不可用状态
工作伙伴	JCheckBox	ChangeEvent	勾选"工作伙伴"复选框则取消勾选"家人"复选框,同时设置"公司名称""地址"等文本框可用;若取消勾选"工作伙伴"复选框,则将"名称""地址"等文本框清空并设置为不可用状态

续表

组件	类型	响应事件	描述
保存	JButton	ActionEvent	接收用户输入的数据
清空	JButton	ActionEvent	清空文本框，重绘对话框
返回	JButton	ActionEvent	关闭对话框

1. "编辑（添加）联系人"对话框与主界面之间的参数传递

主界面 PhoneBookGUI 与"编辑（添加）联系人"对话框 ContractDetailGUI 类之间需要传递待编辑（添加）的联系人信息。如果直接采用 Contract 对象作为参数，则会存在一个问题：如果将联系人修改为其子类对象，则无法将子类对象的属性传递到主界面。

基于上述考虑，程序以主界面的引用为参数进行数据传递。一方面，在 ContractDetailGUI 类中通过主界面的 getCurrentContract() 方法获取待修改联系人的引用，将联系人数据由 PhoneBookGUI 传递到 ContractDetailGUI 类；另一方面，在编辑结束后可以调用 PhoneBookGUI 的 setCurrentContract() 方法，将编辑后的联系人对象赋值给主界面的联系人对象。

2. "编辑（添加）联系人"对话框的初始化

在 ContractDetailGUI 类中定义 initComponent() 方法，实现如下功能：
(1) 实例化对话框的各种组件。
(2) 根据 Contract 对象的属性依次对各组件赋值。
(3) 定义各组件的事件监听器。
initComponent() 方法的详细代码如下。

```java
private void initComponent() {   //实例化对话框的各种组件
    btnAdd = new JButton("添加电话");
    tfdName = new JTextField(10);
    tfdGender = new JTextField(10);
    tfdEmail = new JTextField(15);
    tfdPhones = new ArrayList<JTextField>();
    chkFamily = new JCheckBox("家人", false);
    chkPartner = new JCheckBox("工作伙伴", false);
    tfdAddr = new JTextField(20);
    tfdBirth = new JTextField(10);
    setFamilyEnable(false);
    title = new JTextField(5);
    companyName = new JTextField(10);
    companyPhone = new JTextField(12);
    companyAddr = new JTextField(20);
    companyFax = new JTextField(12);
    setPartnerEnable(false);
```

```java
popupMenu = new JPopupMenu();
del = new JMenuItem("删除电话号码");
popupMenu.add(del);
//根据Contract对象的属性依次对各组件赋值
tfdName.setText(contract.getName() == null ||
    contract.getName().equals("") ?"姓名" : contract.getName());
tfdGender.setText(contract.getGender() == null ||
    contract.getGender().equals("") ?"性别" : contract.getGender());
tfdEmail.setText(contract.getEmail() == null ||
    contract.getEmail().equals("") ?"电子邮件" : contract.getEmail());
if (contract.getPhones() != null){
    for (String p : contract.getPhones())
        tfdPhones.add(createPhoneField(p, 12));
}else
    tfdPhones.add(createPhoneField("", 12));
palPhone = new Box(BoxLayout.Y_AXIS);
for (JTextField phone : tfdPhones)
    palPhone.add(phone);
if (contract instanceof Family) {            //家人
    chkFamily.setSelected(true);
    Family f = (Family) contract;
    tfdAddr.setText(f.getAddress());
    if(f.getBirthday() != null) {
        SimpleDateFormat sdf = new SimpleDateFormat("yyyy-MM-dd");
        tfdBirth.setText(sdf.format(f.getBirthday()));
    }
    setFamilyEnable(true);
}
if (contract instanceof Partner) {           //工作伙伴
    chkPartner.setSelected(true);
    Partner p = (Partner) contract;
    title.setText(p.getTitle());
    Company com = p.getCompany();
    if (com != null) {
        companyName.setText(com.getName());
        companyPhone.setText(com.getPhone());
        companyAddr.setText(com.getAddress());
        companyFax.setText(com.getFax());
    }
    setPartnerEnable(true);
}
chkFamily.addChangeListener(new ChangeListener() {       //定义各组件的事件监听器
    @Override
    public void stateChanged(ChangeEvent e) {
```

```java
            if (chkFamily.isSelected()) {
                chkPartner.setSelected(false);
                setFamilyEnable(true);
            } else {
                tfdAddr.setText("");
                tfdBirth.setText("");
                setFamilyEnable(false);
            }
        }
    });
    chkPartner.addChangeListener(new ChangeListener() {
        @Override
        public void stateChanged(ChangeEvent e) {
            if (chkPartner.isSelected()) {
                chkFamily.setSelected(false);
                setPartnerEnable(true);
            } else {
                title.setText("");
                companyName.setText("");
                companyPhone.setText("");
                companyAddr.setText("");
                companyFax.setText("");
                setPartnerEnable(false);
            }
        }
    });
    btnAdd.addActionListener(new ActionListener() {
        @Override
        public void actionPerformed(ActionEvent e) {
            JTextField newPhone = createPhoneField("", 12);
            tfdPhones.add(newPhone);
            palPhone.add(newPhone);
            ContractDetailGUI.this.validate();      //添加组件之后必须要确认
        }
    });
    del.addActionListener(new ActionListener() {
        @Override
        public void actionPerformed(ActionEvent e) {
            JMenuItem m = (JMenuItem) e.getSource();
            JPopupMenu p = (JPopupMenu) m.getParent();       //获取 JPopupMenu 对象
            JTextField phone = (JTextField) p.getInvoker();  //获取 JTextField 对象
            tfdPhones.remove(phone);
            palPhone.remove(phone);
            ContractDetailGUI.this.repaint();       //刷新界面，删除组件可以仅重绘
        }
```

```
        });
}//initComponent()方法定义结束
```

3. 定义 createPhoneField() 方法

在联系人信息界面中,用户单击"添加电话"按钮后需要动态生成 JTextFiled 组件,并为生成的 JTextFiled 组件设置 MouseEvent 事件处理器。所以需要定义 createPhoneField() 方法,提供上述功能。

```java
private JTextField createPhoneField(String text, int length) {
    JTextField newPhone = new JTextField(text, length);
    newPhone.addMouseListener(new MouseAdapter() {
        public void mouseReleased(MouseEvent e) {
            if (e.isPopupTrigger())           //若释放鼠标右键
                popupMenu.show(e.getComponent(), e.getX(), e.getY());
        }
    });
    return newPhone;
}
```

4. 定义 setFamilyEnable() 方法与 setPartnerEnable() 方法

```java
private void setFamilyEnable(boolean isEnable) {
    tfdAddr.setEnabled(isEnable);
    tfdBirth.setEnabled(isEnable);
}
private void setPartnerEnable(boolean isEnable) {
    title.setEnabled(isEnable);
    companyName.setEnabled(isEnable);
    companyAddr.setEnabled(isEnable);
    companyPhone.setEnabled(isEnable);
    companyFax.setEnabled(isEnable);
}
```

5. 定义 ContractDetailGUI 类的构造方法

```java
public ContractDetailGUI(PhoneBookGUI owner) {
    super(owner, true);
    contract = owner.getCurrentContract();
    initComponent();
    Box base = new Box(BoxLayout.Y_AXIS);              //垂直方向的盒子
    TitledBorder border = BorderFactory.createTitledBorder("基本信息");
    base.setBorder(border);
    base.add(tfdName);
    base.add(tfdGender);
    base.add(tfdEmail);
```

```java
base.add(btnAdd);
base.add(palPhone);
Box family = new Box(BoxLayout.Y_AXIS);          //家人
family.setBorder(BorderFactory.createLineBorder(Color.DARK_GRAY));
family.add(chkFamily);
family.add(new JLabel("地址"));
family.add(tfdAddr);
family.add(new JLabel("出生年月日"));
family.add(tfdBirth);
Box partner = new Box(BoxLayout.Y_AXIS);         //工作伙伴
partner.setBorder(BorderFactory.createLineBorder(Color.DARK_GRAY));
partner.add(chkPartner);
partner.add(new JLabel("职务/职称"));
partner.add(title);
partner.add(new JLabel("所在公司信息:"));
partner.add(new JLabel("名称"));
partner.add(companyName);
partner.add(new JLabel("地址"));
partner.add(companyAddr);
partner.add(new JLabel("电话"));
partner.add(companyPhone);
partner.add(new JLabel("传真"));
partner.add(companyFax);
JPanel palCmd = new JPanel();                    //命令面板
palCmd.setBorder(BorderFactory.createRaisedBevelBorder());
JButton btnSave = new JButton("保存");
JButton btnClear = new JButton("清空");
JButton btnRtn = new JButton("返回");
palCmd.add(btnSave);
palCmd.add(btnClear);
palCmd.add(btnRtn);
Container c = this.getContentPane();
c.setLayout(new BoxLayout(c, BoxLayout.Y_AXIS));
c.add(base);
c.add(family);
c.add(partner);
c.add(palCmd);
setSize(300, 850);
setVisible(true);
btnRtn.addActionListener(new ActionListener() {    //返回按钮
    @Override
    public void actionPerformed(ActionEvent e) {
        int choice = JOptionPane.showConfirmDialog(
                        ContractDetailGUI.this, "确认返回操作吗?");
```

```java
            if (JOptionPane.YES_OPTION == choice)
                ContractDetailGUI.this.dispose();
        }
    });
    btnClear.addActionListener(new ActionListener() {    //清除按钮
        @Override
        public void actionPerformed(ActionEvent e) {
            tfdName.setText("");
            tfdGender.setText("");
            tfdEmail.setText("");
            tfdPhones = new ArrayList<JTextField>();
            tfdPhones.add(createPhoneField("",12));
            palPhone.removeAll();
            palPhone.add(tfdPhones.get(0));
            chkFamily.setSelected(false);
            chkPartner.setSelected(false);
            ContractDetailGUI.this.validate();
        }
    });
    btnSave.addActionListener(new ActionListener() {    //保存按钮
        @Override
        public void actionPerformed(ActionEvent e) {
            String name = tfdName.getText().equals("姓名") ? ""
                                    : tfdName.getText();
            String gender = tfdGender.getText().equals("性别") ? ""
                                    : tfdGender.getText();
            String email = tfdEmail.getText().equals("电子邮件") ? ""
                                    : tfdEmail.getText();
            Set<String> phones = new HashSet<String>();
            for (JTextField p : tfdPhones) {
                phones.add(p.getText());
            }
            if (chkFamily.isSelected()) {                    //家人
                java.util.Date birthday = null;
                if (!tfdBirth.getText().equals("")) {
                    SimpleDateFormat sdf = new SimpleDateFormat(
                                        "yyyy-MM-dd");
                    try {
                            birthday = sdf.parse(tfdBirth.getText());
                        } catch (ParseException e1) {
                            e1.printStackTrace();
                        }
                }
                Family f = null;
```

```java
            try {
                f = new Family(name, gender, email, phones, birthday,
                        tfdAddr.getText());
            } catch (Exception e1) {
                JOptionPane.showMessageDialog(ContractDetailGUI.this,
                        e1.getMessage());
            }
            owner.setCurrentContract(f);        //将家人信息传回主界面
            return;
        }
        if (chkPartner.isSelected()) {          //工作伙伴
            Partner p = null;
            try {
                Company t = new Company(companyName.getText(),
                    companyAddr.getText(), companyPhone.getText(),
                    companyFax.getText())
                p = new Partner(name, gender, email, phones,
                        title.getText(),t);
            } catch (Exception e1) {
                JOptionPane.showMessageDialog (ContractDetailGUI.this,
                        e1.getMessage());
            }
            owner.setCurrentContract(p);        //将工作伙伴信息传回主界面
            return;
        }
        Contract c = null;                      //一般联系人
        try {
            c = new Contract(name, gender, email, phones);
        } catch (Exception e1) {
            JOptionPane.showMessageDialog(ContractDetailGUI.this,
                    e1.getMessage());
        }
        owner.setCurrentContract(c);            //将联系人信息传回主界面
        return;
        }
    });
}//ContractDetailGUI 类构造方法结束
```

10.4.4 通讯录业务逻辑类 PhoneBook

10.4.4

通讯录业务逻辑 PhoneBook 类为图形用户界面 cn.edu.sdtbu.phoneBook.gui 层提供服务,其具体实现如下。

1. 定义 PhoneBook 类构造方法

PhoneBook 类封装联系人对象列表，实现联系人按照姓名拼音升序排列。

```java
public class PhoneBook {
    private List<Contract> contracts;
    public void setContracts(List<Contract> contracts) {
        contracts = contracts;
        Collections.sort(contracts);           //有序列表
    }
    public PhoneBook(){
        contracts = new ArrayList<Contract>();
    }
    public PhoneBook(List<Contract> contracts) {
        setContracts(contracts);               //调用 setter 方法,保证联系人有序
    }
```

2. 定义显示、删除、模糊查询、添加、修改等操作

```java
public void display(){                         //显示联系人列表
    for(Contract c:getContracts())
        c.display();
}
public boolean deleteContract(Contract c) {    //删除联系人
    return contracts.remove(c);
}
public boolean updateContract(Contract c){     //修改联系人信息
    int index = Collections.binarySearch(getContracts(), c);
    if(index != -1)   {
        contracts.set(index,c);                     ①
        return true;
    }else
        return false;
}
public List<Contract> findContainsByName(String name) {    //模糊查询
    List<Contract> result = new ArrayList<Contract> ();
    for (Contract c:getContracts()) {
        if (c.getName().contains(name))
            result.add(c);
    }
    return result;
}
public void clearContracts() {                 //清空
    contracts = null;
}
```

```java
public void addContract(Contract c){
    if(contracts == null){                        //通讯录为空
        contracts = new ArrayList<Contract>();
        contracts.add(c);
        return;
    }
    int index = Collections.binarySearch(getContracts(), c);
    if(index < 0){                                //没有找到
        getContracts().add(-index-1, c);          //插入联系人,并保证升序排列
    }else{                                        //找到则合并
        getContracts().get(index).mergeContract(c);
    }
}
}//PhoneBook类定义结束
```

上述代码中的 updateContract() 方法实现了修改联系人信息,其中,代码行①利用 JList 列表提供的 set() 方法实现了联系人的引用替换。

注意:此处不可以采用如下实现方式。

```java
try {
    contracts.get(index).update(c);
} catch (Exception e) {
    JOptionPane.showMessageDialog(null, e.getMessage());
}
```

主要原因在于待修改联系人的对象类型可能会发生变化,如将原来的 Contract 类型的联系人变更为 Family 类型,根据多态原则,此处只能调用 Contract 类型的 update() 方法,这样就会遗失对象 c 中关于 Family 类型的属性。

10.5 练一练

1. 个人通讯录(五)中的删除联系人功能每次仅能删除一人,如何实现同时删除多个联系人的操作?

提示:查阅 Java API 文档,学习设置 JList 组件的选择模式的方法。

2. 编程实现基于 GUI 的 100 以内的加减法生成器。

要求:随机生成 100 以内的加减法运算,其中减法运算的结果不能小于 0。

3. 采用与个人通讯录(五)类似的系统架构实现系统的登录操作。

(1) 实体类 User。

描述用户信息,包含 Id、昵称、密码等信息。其中,昵称不能为空,长度为小于或等于 8 个字符的字符串;密码不能为空,长度为 6~8 个字符的字符串。

(2) 业务逻辑类 UserService。

封装用户列表,提供登录方法。

第 11 章　JDBC 编程——个人通讯录(六)

11.1　个人通讯录(六)

在个人通讯录(五)的基础上设计与实现基于 MySQL 的个人通讯录系统,要求:
(1) 构建具有分层架构的个人通讯录系统;
(2) 利用 MySQL 数据库实现通讯录数据的持久化存储。

11.2　程序设计思路

个人通讯录(六)基于分层架构,所谓分层架构就是指将系统拆分成若干层,每层都有清晰的角色和分工,下层为上层提供服务。每层不需要关心其他层的实现细节,通过使用接口定义该层的功能逻辑,这将允许创建该接口的不同实现,很容易实现用新的实现替换原有层次。

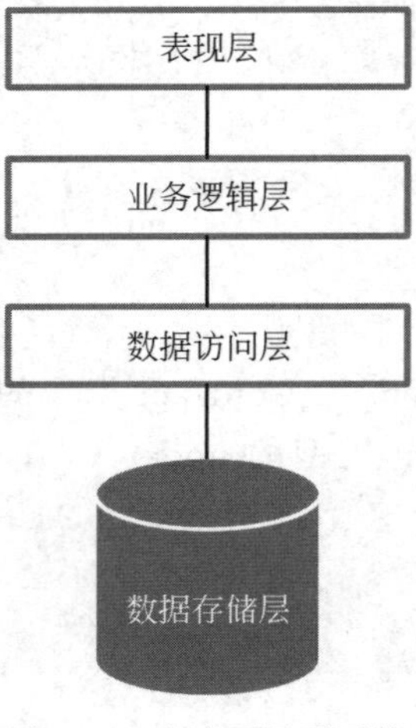

分层架构降低了层与层之间的依赖,是一种高内聚、低耦合的设计模式。分层架构使开发人员的分工更明确,能够更专注于应用系统核心业务逻辑的分析、设计和开发工作。

个人通讯录系统可以分为以下 4 层,如图 11-1 所示。

表现层:主要指用户界面,用来展示功能,实现与用户的交互。

业务逻辑层:实现系统的业务逻辑,如添加联系人、删除联系人等。

数据访问层:实现对数据表的增、删、改、查等操作。

数据存储层:利用 MySQL 数据库实现通讯录数据的持久化存储。

图 11-1　个人通讯录系统的分层架构

11.3　关键技术

11.3.1　JDBC 简介

JDBC(Java DataBase Connection)的全称为 Java 数据库连接,它是面向应用程序开发人员和数据库驱动程序开发人员的应用程序接口。通过 JDBC,开发人员可以很方便地操作各种主流数据库,用 JDBC 编写的程序能够自动将 SQL 语句传送给相应的数据库管理系统(DBMS),从而真正实现跨平台编程。

11.3.2 MySQL 的安装

MySQL 的体积小、速度快且开放源代码，是中小型网站开发的首选数据库。

登录 MySQL 官方网站（https://dev.mysql.com/downloads/installer/）下载最新版本的安装程序，本书定稿之时，最新的版本是 MySQL 8.0，下载的是安装文件 mysql-installer-web-community-8.0.13.0.msi。无论是 32 位系统还是 64 位系统都可以使用该安装程序进行 MySQL 的安装。

安装过程可以选择自定义安装方式，也可以选择 MySQL Server、Connector J 进行安装。安装之前，安装文件会首先检测系统环境是否满足安装要求，如果不满足，则会给出提示及相应的解决方案；如果满足条件，则会开始下载和安装。

安装成功之后，安装文件会引导用户进行系统配置，重新定义或者采用默认的端口号 3306，设置 root 用户的密码，此处可以添加新的数据库用户。

11.3.3 JDBC 常用接口及类

Java 9 及以上版本开始支持 JDBC 4.3 标准，提供如下常用接口及类定义。

1. DriverManager 类

JDBC 驱动程序管理器，定义获取连接的方法为

```
public static Connection getConnection(String url,String user,String password)
                                                       throws SQLException
```

其中，参数 url 对应数据库的 URL 描述信息，user 和 password 分别表示用户名与密码。访问数据库之前，必须首先与数据库建立连接。

2. Connection 接口

描述与数据库连接的对象。Connection 接口定义了如下常用方法。

```
Statement createStatement() throws SQLException
```

创建 Statement 对象。

```
PreparedStatement prepareStatement(String sql) throws SQLException
```

创建预编译的 Statement 对象，PreparedStatement 是 Statement 的子接口。

3. Statement 接口

执行 SQL 语句的工具接口，常用方法如下。

```
ResultSet executeQuery(String sql) throws SQLException
```

执行 SQL 查询语句,返回满足条件的结果集 ResultSet 对象。

```
int executeUpdate(String sql) throws SQLException
```

若执行 update、insert 或者 delete 语句,则返回受影响的记录条数(也称更新计数);若执行 create table 或 drop table 等语句,则 executeUpdate()的返回值是零。

4. PreparedStatement 接口

Statement 的子接口,表示预编译的 Statement 对象。提供对 SQL 语句的预编译处理,通常预编译的 SQL 语句中包含"?"表示的参数。

除了继承父接口 Statement 中的方法之外,还定义了如下方法为预编译的 SQL 语句传递参数。

```
void setXXX(int parameterIndex, XXX x) throws SQLException
```

XXX 指定某种数据类型,parameterIndex 表示参数的索引号。

```
void setNull(int parameterIndex, int sqlType) throws SQLException
```

将指定参数设置为 null。parameterIndex 表示参数的索引号,sqlType 表示参数字段的类型。

5. ResultSet 接口

表示查询结果集对象,接口中定义的获取查询结果的方法为

```
XXX getXXX(int columnIndex) throws SQLException
XXX getXXX(String columnLabel) throws SQLException
```

获取当前记录指定列的值,列可以通过索引(columnIndex)或者列名称(columnLabel)确定。

ResultSet 接口中提供了许多改变当前记录的方法,如:

```
boolean next() throws SQLException
```

将当前记录定位于下一行。

11.3.4 JDBC 操作数据库的基本步骤

应用程序中,通过 JDBC 操作数据库通常按照如下步骤执行。

1. 加载数据库驱动

利用 java.lang.Class 类提供的 forName()方法加载数据库驱动到 JVM。

```
public static Class<?> forName(String className) throws ClassNotFoundException
```

参数 className 表示数据库驱动的全限定名称。例如,加载 MySQL 数据库驱动的

代码为

```
Class.forName("com.mysql.cj.jdbc.Driver");
```

注意：MySQL 5 之前的版本，其数据库驱动名称为 com.mysql.jdbc.Driver；从 MySQL 6.0 开始，驱动更名为 com.mysql.cj.jdbc.Driver。

2. 获取数据库连接

利用 DriverManager 类连接数据库获取 Connection 连接对象。例如：

```
Connection Conn = DriverManager.getConnection("jdbc:mysql://localhost:3306/book? useSSL = false & serverTimezone=UTC","用户名","密码");
```

通过上述命令与 MySQL 数据库 book 建立连接，其中，localhost 是数据库所在 IP 地址，3306 为端口号，"用户名""密码"使用配置 MySQL 时创建的用户及密码。从 MySQL 6.0 开始，与数据库建立连接时需要明确指明服务器端的时区，上述连接中使用 UTC (Universal Time Coordinated)时区。另外，本系统不需要 SSL 连接，所以连接选项中显式禁用 SSL 连接。

3. 创建 Statement 对象或 PrepareStatement 对象

通过 Connection 对象创建 Statement 或 PreparedStatement 对象。

4. 执行 SQL 语句

利用 Statement 对象或 PreparedStatement 对象调用 executeQuery()方法或 executeUpdate()方法执行 SQL 语句。

5. 处理结果集

对于查询操作，需要从 ResultSet 中获取满足条件的记录并将其封装在集合中；对于其他更新操作，则根据 executeUpdate()方法的返回值判定 SQL 语句的执行情况。

6. 回收数据库资源

依次关闭 ResultSet 对象、Statement 对象或者 PreparedStatement 对象及 Connection 对象。

例 11-1 示范了通过 JDBC 查询数据库的过程。

例 11-1

```java
public class JDBCTest {
    public static void main(String[] args)
                            throws ClassNotFoundException, SQLException {
        Class.forName("com.mysql.cj.jdbc.Driver");
        Connection conn = DriverManager.getConnection(
            "jdbc:mysql://localhost:3306/sprinttest? useSSL= false
```

```
                    &serverTimezone= UTC","yulp","123456");
Statement s = conn.createStatement();
ResultSet rs = s.executeQuery("select uid,uname,usex from user");
while(rs.next()) {
    int uid = rs.getInt(1);
    String uname = rs.getString(2);
    String usex = rs.getString(3);
    System.out.println("uid:"+uid+",uname:"+uname+",usex:"+usex);
}
System.out.println("----------");
PreparedStatement ps = conn.prepareStatement("select
                    uid,uname,usex from user where usex=?");
ps.setString(1, "女");
rs = ps.executeQuery();
while(rs.next()) {
    int uid = rs.getInt(1);
    String uname = rs.getString(2);
    String usex = rs.getString(3);
    System.out.println("uid:"+uid+",uname:"+uname+",usex:"+usex);
}
rs.close();
s.close();
ps.close();
conn.close();
    }
}
```

例 11-1 的运行结果如图 11-2 所示。

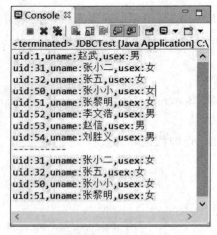

图 11-2　例 11-1 的运行结果

11.4 个人通讯录（六）设计步骤

11.4.1 数据库设计

1. 数据库逻辑设计

根据项目需求对数据库进行逻辑设计,得到如图 11-3 所示的实体-关系(ER)图,此处假设联系人最多在一家公司就职。

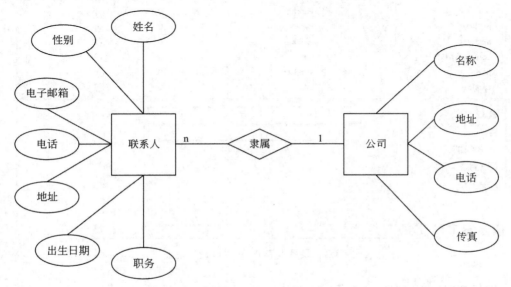

图 11-3　个人通讯录（六）实体-关系图

2. 创建数据库

根据 ER 图,在 MySQL 系统中使用 create table 语句设计联系人表(Contracts)及公司表(Companies)。关于数据库的操作,可以通过 Navicat for MySQL 客户端以图形化的方式执行,操作步骤如下。

(1) 创建新的连接 phoneBook。

注意：Navicat 创建连接时需要提供登录用户名及密码。

(2) 新建数据库 book 及数据表。

在 Navicat 中选中连接 phoneBook,单击右键,创建新的数据库 book。在数据库 book 中创建联系人表 Contracts,设计界面如图 11-4 所示。

其中,数据表中定义自动增长的主键 id,使数据表可以存储同名的联系人信息,添加一项 catalog,用于标记联系人的类别：普通联系人、家人、工作伙伴。

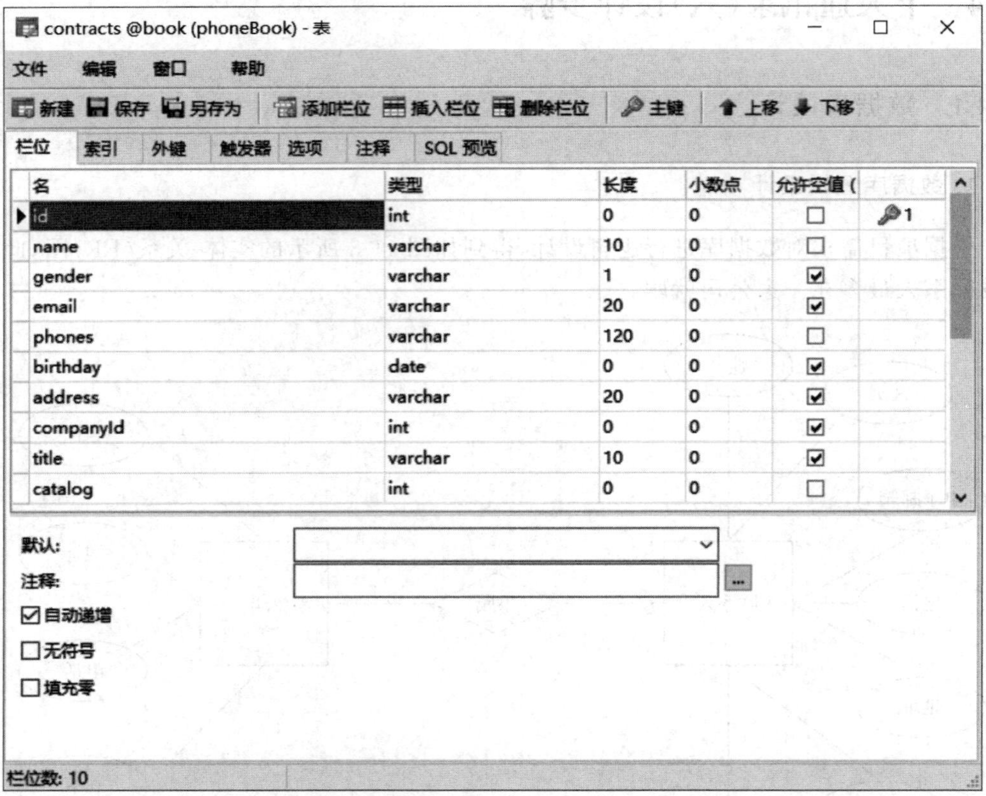

图 11-4　联系人表 Contracts 设计视图

创建外键约束,如图 11-5 所示。

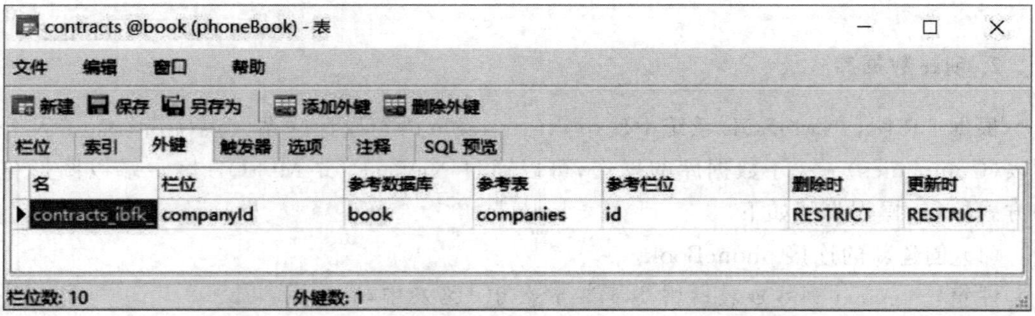

图 11-5　联系人表 Contracts 外键视图

公司表 Companies 的设计界面如图 11-6 所示。其中,数据表的主键 id 也定义为自动增长。

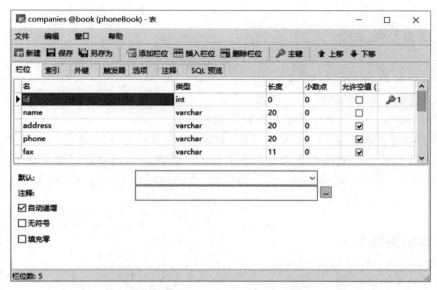

图 11-6 公司表 Companies 设计视图

11.4.2 导入 MySQL 数据库驱动

项目基于 JDBC 访问 MySQL 数据库，需要 MySQL 驱动的支持。实验中使用的类库 mysql-connector-java-8.0.20.jar 位于 connector J 的安装目录下。在 Eclipse 中选择 project/properties/Java Build Path/Liberaies 选项卡下的 Add External JARs 命令，添加 MySQL 驱动，如图 11-7 所示。

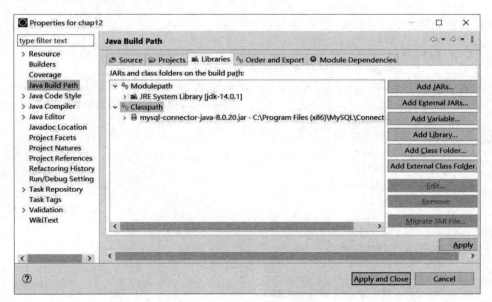

图 11-7 导入 MySQL 驱动

11.4.3 项目总体结构

本项目分为如下几层。
（1）GUI 层：系统的图形用户界面。
（2）Service 层：实现通讯录系统的业务逻辑。
（3）DAO 层：通过 JDBC 连接 MySQL 数据库，实现数据表的增、删、改、查操作。

在个人通讯录（五）的基础上增加两个数据包：数据访问包 dao 可以实现对 book 数据库的 Contracts 及 Companies 数据表的增、删、改、查操作；工具包 util 封装了数据库工具类。

11.4.4

11.4.4 数据实体 bean

1. 联系人 Contract 类

Contract 类描述联系人信息，在个人通讯录（五）的基础上增加了成员变量 id，通过 id 唯一标识联系人对象。本程序中，Contract 对象的排序是由数据库的排序操作实现的，所以此处定义的 Contract 类无须实现 Comparable<>接口。联系人 Contract 类的核心代码如下所示。

```
public class Contract{
    private int id;                    //编号
    private String name;               //姓名
    private String gender;             //性别
    private String email;              //电子邮箱
    private Set<String> phones;        //电话
    public int getId() {
        return id;
    }
    public Contract(int id,String name, String gender, String email,
                                    Set< String> phones) throws Exception {
        this.id = id;
        setName(name);
        setGender(gender);
        setEmail(email);
        setPhones(phones);
    }
}
```

2. 家庭联系人 Family 类

```
public class Family extends Contract {
    //省略与个人通讯录(五)相同的代码
    public Family(int id,String name, String gender, String email,
```

```
        Set<String> phones,Date birthday, String address) throws Exception {
            super(id,name, gender, email, phones);
            setBirthday(birthday);
            setAddress(address);
        }
    }
```

3. 工作伙伴 Partner 类

```
public class Partner extends Contract {
    //省略与个人通讯录(五)相同的代码
    public Partner(int id,String name, String gender, String email,
        Set<String> phones, String title, Company company) throws Exception {
            super(id,name, gender, email, phones);
            setCompany(company);
            setTitle(title);
        }
    }
```

4. 公司 Company 类

```
public class Company {
    private int id;              //增加公司编号
    //省略与个人通讯录(五)相同的代码
    public Company(int id,String name, String address, String phone,
                    String fax) {
        setId(id);
        setName(name);
        setAddress(address);
        setPhone(phone);
        setFax(fax);
    }
}
```

11.4.5 工具包 util

定义类 DBTool,封装与远程数据库的连接对象,提供获取与关闭数据库连接的方法。

11.4.5

```
public class DBTool {
    private static Connection conn = null;
    public static Connection getConnection(){      //获取连接
        try {
            if(conn == null || conn.isClosed()){
                try {
```

```java
            Class.forName("com.mysql.cj.jdbc.Driver");
            conn=DriverManager.getConnection("jdbc:mysql://localhost:
        3306/book?useSSL=false&serverTimezone=UTC","yulp","123456");
            } catch (ClassNotFoundException e) {
                e.printStackTrace();
            } catch (SQLException e) {
                e.printStackTrace();
            }
        }
    } catch (SQLException e) {
        e.printStackTrace();
    }
    return conn;
}
public static void closeConnection(){       //关闭连接
    try {
        if(conn != null && !conn.isClosed()){
            conn.close();
            conn = null;
        }
     }catch (SQLException e) {
        e.printStackTrace();
    }
}
}//DBTool类定义结束
```

11.4.6 数据访问包 dao

11.4.6

本项目涉及的数据访问对象包括联系人表 Contracts 和联系人所在的公司表 Companies，dao 层分别定义接口及类实现，以对 Contracts 及 Companies 数据表进行增、删、改、查操作。dao 层对异常信息仅是简单抛出，不进行处理。

1. ContractDao 接口

```java
public interface ContractDao {
    boolean add(Contract c) throws Exception;
    boolean deleteById(int contractId) throws Exception;
    boolean update(Contract c) throws Exception;
    List<Contract> searchByName(String name) throws Exception;
}
```

2. ContractDaoImpl 类

（1）ContractDaoImpl 类的定义。

```
public class ContractDaoImpl implements ContractDao {
    //定义 getStringPhones 方法转换电话号码
    //实现接口的 searchByName()方法
    //实现接口的 add()方法
    //实现接口的 update()方法
    //实现接口的 deleteById()方法
}
```

(2) 转换电话号码。

定义私有方法 getStringPhones(Set<String> phones)，实现将以集合 Set 形式存储的若干电话号码转换为字符串类型，其中，电话号码之间用分号连接。

```
private String getStringPhones(Set<String> phones) {
    if (phones.size() == 0)
        return null;
    StringBuilder sb = new StringBuilder();
    for (String str : phones) {
        sb.append(str);
        sb.append(";");
    }
    return sb.substring(0, sb.length() - 1);
}
```

(3) 查询联系人。

根据用户输入的联系人姓名进行模糊查询，返回满足条件的联系人列表；如果输入空串，则获取所有的联系人信息。查询联系人操作需要解决以下 3 个问题。

① 对子类 Partner 对象，如何将 Contracts 表与 Companies 表进行合并？

解决方法：将 Contracts 数据表左连接 Companies 数据表，在保留联系人表所有记录的基础上与 Companies 表进行合并，提取公司信息。

② 如何进行模糊查询？

解决方法：使用 SQL 语句的 like 短语及通配符"?"进行模糊查询。

③ 查询结果如何按照联系人姓名的升序排列？

解决方法：利用 MySQL 数据库定义的 CONVERT（表达式 USING 编码方式）短语将表达式编码为指定编码格式（GBK），再使用 order by 短语对满足条件的联系人进行排序。

具体实现代码如下。

```
public List<Contract> searchByName(String name) throws Exception {
    List<Contract> list = new ArrayList<Contract>();
    Connection conn = DBTool.getConnection();
    ResultSet rs;
    PreparedStatement pst = conn.prepareStatement("select
        contracts.*,companies.* from contracts left join companies on
```

```java
            contracts.companyId= companies.id where contracts.name like ? order
            by convert(contracts.name using GBK)");
    pst.setString(1, "%" + name + "%");
    rs = pst.executeQuery();
    while (rs.next()) {
        int id = rs.getInt(1);
        String n = rs.getString(2);
        String gender = rs.getString(3);
        String email = rs.getString(4);
        String phones = rs.getString(5);
        Date birth = rs.getDate(6);
        String addr = rs.getString(7);
        int companyId = rs.getInt(8);
        String title = rs.getString(9);
        int catalog = rs.getInt(10);
        String cName = rs.getString(12);
        String cAddr = rs.getString(13);
        String cPhone = rs.getString(14);
        String cFax = rs.getString(15);
        Set<String> listPhones = new HashSet<String> (Arrays.asList(
                            phones.split(";")));
        if (catalog == 1) {
            Family f = new Family(id, n, gender, email, listPhones,
                birth== null?null:new java.util.Date(birth.getTime()),
                addr);
            list.add(f);
        } else if (catalog == 2) {
            Partner p = new Partner(id, n, gender, email, listPhones,
            title, new Company(companyId,cName,cAddr,cPhone,cFax));
            list.add(p);
        } else {
            Contract c = new Contract(id,n,gender,email,listPhones);
            list.add(c);
        }
    }
    rs.close();
    pst.close();
    return list;
}
```

（4）添加联系人。

将联系人或子类实体对象添加到数据表 Contracts，添加成功则返回 true，否则返回 false。

```java
public boolean add(Contract c) throws SQLException {
```

```java
    boolean flag = false;
    String strPhones = getStringPhones(c.getPhones());
    int catalog = 0;
    java.sql.Date birthday = null;
    String address = null;
    int companyId = 0;
    String title = null;
    if (c.getClass() == Family.class) {            //家人
        catalog = 1;
        java.util.Date birth = ((Family) c).getBirthday();
        if (birth != null)
            birthday = new java.sql.Date(birth.getTime());
        address = ((Family) c).getAddress();
    }else if (c.getClass() == Partner.class) {     //工作伙伴
        catalog = 2;
        companyId = ((Partner) c).getCompany().getId();
        title = ((Partner) c).getTitle();
    }
    Connection conn = DBTool.getConnection();
    String sql = "insert into contracts(name,gender,email,
                 phones,birthday,address,companyId,title,catalog)
                 values(?,?,?,?,?,?,?,?,?)";
    PreparedStatement state = conn.prepareStatement(sql);
    state.setString(1, c.getName());
    state.setString(2, c.getGender());
    state.setString(3, c.getEmail());
    state.setString(4, strPhones);
    state.setDate(5, birthday);
    state.setString(6, address);
    if (companyId == 0)
        state.setNull(7, Types.INTEGER);
    else
        state.setInt(7, companyId);
    state.setString(8, title);
    state.setInt(9, catalog);
    int result = state.executeUpdate();
    if (result == 1) {
        flag = true;
    }
    state.close();
    return flag;
}
```

(5) 修改联系人。

修改指定编号(id)的联系人信息,修改成功则返回 true,否则返回 false。

```java
public boolean update(Contract c) throws SQLException {
    boolean flag = false;
    String strPhones = getStringPhones(c.getPhones());
    int catalog = 0;
    String address = null;
    java.sql.Date birthday = null;
    int companyId = 0;
    String title = null;
    if (c.getClass() == Family.class) {            //家人
        catalog = 1;
        java.util.Date birth = ((Family) c).getBirthday();
        if (birth != null)
            birthday = new java.sql.Date(birth.getTime());
        address = ((Family) c).getAddress();
    }else if (c.getClass() == Partner.class) {    //工作伙伴
        catalog = 2;
        companyId = ((Partner) c).getCompany().getId();
        title = ((Partner) c).getTitle();
    }
    Connection conn = DBTool.getConnection();
    String sql = "update contracts set name=?,gender=?,phones=?,
        email=?,birthday=?,address=?,companyId=?,title=?,catalog=? where
        id=?";
    PreparedStatement state = conn.prepareStatement(sql);
    state.setString(1, c.getName());
    state.setString(2, c.getGender());
    state.setString(3, strPhones);
    state.setString(4, c.getEmail());
    state.setDate(5, birthday);
    state.setString(6, address);
    if (companyId == 0)
        state.setNull(7, Types.INTEGER);
    else
        state.setInt(7, companyId);
    state.setString(8, title);
    state.setInt(9, catalog);
    state.setInt(10, c.getId());
    int result = state.executeUpdate();
    if (result == 1) {
        flag = true;
    }
```

```
        state.close();
        return flag;
}
```

（6）删除联系人。

删除指定编号的联系人记录，删除成功则返回 true，否则返回 false。

```
@Override
    public boolean deleteById(int contractId) throws SQLException {
        boolean flag = false;
        Connection conn = DBTool.getConnection();
        Statement state = conn.createStatement();
        String sql = "delete from contracts where id =" + contractId;
        int result = state.executeUpdate(sql);
        if (result == 1)
            flag = true;
        state.close();
        return flag;
    }
```

3. CompanyDao 接口

```
public interface CompanyDao {
    int add(Company c) throws Exception;
    boolean update(Company c) throws Exception;
    List<Company> searchAll() throws Exception;
}
```

4. CompanyDaoImpl 类

（1）CompanyDaoImpl 类的定义。

```
public class CompanyDaoImpl implements CompanyDao {
    //实现接口的 add()方法
    //实现接口的 update()方法
    //实现接口的 searchAll()方法
}
```

（2）添加公司。

将公司实体对象添加到数据表 Companies，添加成功则返回刚刚插入记录的 id 字段值，否则返回-1。

获取插入记录的自增主键，可以通过 Statement 接口的下列方法实现：

```
int executeUpdate(String sql, Statement.RETURN_GENERATED_KEYS) throws SQLException
```

执行 SQL 语句，并生成可查询的自增主键。

```
ResultSet getGeneratedKeys() throws SQLException
```

返回刚刚生成的自增主键。

程序的具体实现如下。

```java
public int add(Company c) throws Exception {
    int key = -1;
    Connection conn = DBTool.getConnection();
    Statement state = conn.createStatement();
    String sql = "insert into companies(name,address,phone,fax)
                  values ("'"+c.getName()+"','"+c.getAddress()+"','"+
        c.getPhone()+"','"+c.getFax()+"')";
    int result = state.executeUpdate(sql, Statement.RETURN_GENERATED_KEYS);
    if(result == 1) {
        ResultSet keys = state.getGeneratedKeys();
        keys.next();
        key = keys.getInt(1);
    }
    state.close();
    return key;
}
```

(3) 修改公司。

修改指定编号的公司信息,修改成功则返回 true,否则返回 false。

```java
public boolean update(Company c) throws SQLException {
    boolean flag = false;
    Connection conn = DBTool.getConnection();
    String sql = "update companies set name=?,address=?,
        phone=?,fax=? where id=?";
    PreparedStatement state = conn.prepareStatement(sql);
    state.setString(1, c.getName());
    state.setString(2, c.getAddress());
    state.setString(3, c.getPhone());
    state.setString(4, c.getFax());
    state.setInt(5, c.getId());
    int result = state.executeUpdate();
    if(result == 1) {
        flag = true;
    }
    state.close();
    return flag;
}
```

(4) 获取所有公司信息。

```java
public List<Company> searchAll() throws Exception {
    List<Company> result = new ArrayList<Company>();
```

```java
        Connection conn = DBTool.getConnection();
        ResultSet rs;
        Statement st = conn.createStatement();
        rs = st.executeQuery("select * from companies");
        while(rs.next()) {
            result.add(new Company(rs.getInt(1),rs.getString(2),
                rs.getString(3),rs.getString(4),rs.getString(5)));
        }
        rs.close();
        st.close();
        return result;
    }
```

11.4.7 业务逻辑层 service

业务逻辑层通过数据访问层提供的基本操作定义业务规则，实现系统的业务逻辑。

1. 异常处理

数据访问层（dao 层）在进行数据库处理时可能会发生数据库操作异常，抛出 SQLException 等异常信息。对于底层的异常信息，通常不建议直接传到用户界面，一方面，这些异常信息对用户使用系统而言是没有任何帮助的；另一方面，暴露的底层异常信息可能被恶意用户利用，使系统遭受攻击。

通常，业务逻辑层需要捕获数据访问层抛出的异常信息，将捕获的异常信息封装为业务异常再重新抛出。例如本系统定义如下业务异常类。

```java
public class DBException extends Exception {
    public DBException() {
        super("访问数据库发生异常");
    }
}
```

2. ContractService 接口

```java
public interface ContractService {
    List<Contract> searchByName(String name) throws Exception;
    boolean addContract(Contract c) throws Exception;
    boolean deleteContractById(int contractId) throws Exception;
    boolean updateContract(Contract c) throws Exception;
}
```

3. ContractServImpl 类

ContractServImpl 类封装 ContractDao 对象与 CompanyDao 对象，以实现接口

ContractServic 中定义的方法。

(1) ContractServImpl 类的定义。

```java
public class ContractServImpl implements ContractService {
    private ContractDao contractDao;
    private CompanyDao companyDao;
    public ContractServImpl(ContractDao contractDao,CompanyDao companyDao) {
        this.contractDao = contractDao;
        this.companyDao = companyDao;
    }
    //分别实现接口定义的方法
}
```

定义构造方法，实现 contractDao 对象与 companyDao 对象的外部注入。

(2) 查询联系人。

```java
public List<Contract> searchByName(String name) throws Exception {
    try {
        return contractDao.searchByName(name);
    } catch (Exception e) {
        throw new DBException();
    }
}
```

(3) 删除联系人。

```java
public boolean deleteContractById(int contractId) throws Exception {
    try {
        return contractDao.deleteById(contractId);
    } catch (Exception e) {
        throw new DBException();
    }
}
```

(4) 添加联系人。

根据用户输入的联系人或者子类信息，区别 Contract 类、Family 类或者 Partner 类对象并对其分别进行处理。如果联系人是 Contract 类或者 Family 类对象，则直接调用 contractDao 对象的 add()方法，将联系人信息添加到 Contracts 数据表；若联系人是 Partner 类的对象，则首先需要处理公司信息，判断是否有新的公司加入，如果有，则在 Companies 表中添加公司记录，否则修改已存在的公司信息。处理完公司信息之后，再将联系人信息添加到 Contracts 表。

具体实现代码如下。

```java
public boolean addContract(Contract c) throws Exception {
    if(c.getClass() == Contract.class
        || c.getClass() == Family.class) {              //普通联系人或者家人
```

```java
        try {
            return contractDao.add(c);                //添加联系人信息
        } catch (Exception e) {
            throw new DBException();
        }
    }else if(c.getClass() == Partner.class) {         //工作伙伴
        Partner p = (Partner)c;
        int companyId = p.getCompany().getId();
        if(companyId == 0) {                          //添加新的公司
            try {
                companyId = companyDao.add(p.getCompany());
            } catch (Exception e) {
                throw new DBException();
            }
            p.getCompany().setId(companyId);
        }else {                                       //修改公司信息
            try {
                companyDao.update(p.getCompany());
            } catch (Exception e) {
                throw new DBException();
            }
        }
        try {
            return contractDao.add(p);                //添加联系人信息
        } catch (Exception e) {
            throw new DBException();
        }
    }else
        return false;
}
```

(5) 修改联系人。

```java
public boolean updateContract(Contract c) throws Exception {
    if(c.getClass() == Contract.class
        || c.getClass() == Family.class) {            //普通联系人或者家人
        try {
            return contractDao.update(c);             //修改联系人信息
        } catch (Exception e) {
            throw new DBException();
        }
    }else if(c.getClass() == Partner.class) {         //工作伙伴
        Partner p = (Partner)c;
        int companyId = p.getCompany().getId();
        if(companyId == 0) {                          //添加新的公司
```

```java
        try {
            companyId = companyDao.add(p.getCompany());
        } catch (Exception e) {
            throw new DBException();
        }
        p.getCompany().setId(companyId);
    } else {                                          //修改公司信息
        try {
            companyDao.update(p.getCompany());
        } catch (Exception e) {
            throw new DBException();
        }
    }
    try {
        return contractDao.update(p);                 //修改联系人信息
    } catch (Exception e) {
        throw new DBException();
    }
}else
    return false;
}
```

4. CompanyService 接口

```java
public interface CompanyService {
    List<Company> searchAll() throws Exception;
}
```

5. CompanyServImpl 类

封装 CompanyDao 对象，实现 CompanyService 接口中定义的方法。

```java
public class CompanyServImpl implements CompanyService{
    private CompanyDao dao;
    public CompanyServImpl(CompanyDao dao) {
        this.dao = dao;
    }
    @Override
    public List<Company> searchAll() throws Exception {
        try {
            return dao.searchAll();
        } catch (Exception e) {
            throw new DBException();
        }
    }
}
```

11.4.8 图形用户界面层 gui

11.4.8

1. 通讯录主界面

通讯录主界面利用业务逻辑 ContractService 对象为用户提供通讯录的显示、搜索、添加、维护及删除功能。在个人通讯录（五）的基础上修改 ContractService 的定义如下。

```
public class PhoneBookGUI extends JFrame {
    public PhoneBookGUI() {
        ContractDao contractDao = new ContractDaoImpl();
        CompanyDao companyDao = new CompanyDaoImpl();
        ContractService conService = new ContractServImpl(contractDao,
                                                           companyDao);
    }
}
```

2. "编辑(添加)联系人"对话框

"编辑(添加)联系人"对话框主要实现联系人信息的编辑。与个人通讯录（五）的"联系人"对话框基本相同，此处省略其详细代码。

11.5 练一练

1. 讨论系统的分层架构有哪些结构上的优势。
2. 利用分层架构编程实现用户的登录、注册模块。
（1）设计用户数据表 users，存储用户信息：Id、昵称、密码等。
（2）定义实体类 User。
（3）定义数据访问层接口 UserDao、实现类 UserDaoImpl。
（4）定义业务逻辑层接口 UserService、实现类 UserServImpl。
（5）设计与实现登录、注册界面。

第 12 章 输入/输出流——个人通讯录(七)

12.1 个人通讯录(七)

　　个人通讯录(七)为联系人增加了头像属性,即通讯录不仅存储姓名、电话号码等文本信息,还可以存储头像等多媒体信息。用户通过个人通讯录(七)可以上传、下载及展示联系人的头像图片,弥补了通讯录系统千篇一律、过于呆板的不足,增加了个性化设置。

12.2 程序设计思路

　　分析个人通讯录(七)中增加的用户需求,得出本程序的实现关键是客户端到服务器端的多媒体文件的上传和下载问题。

　　文件上传和下载是 Java 应用程序中常见的输入/输出操作。Java 定义了输入/输出数据流的抽象不同来源、不同目标的数据,并提供了输入流的读入操作与输出流的写出操作。文件上传就是将文件通过输入/输出流传送到服务器端;而文件下载则与文件上传相反,是将文件从服务器端通过输入/输出流下载到本地。

12.3 关键技术

12.3.1 File 类

　　Java.io.File 类是磁盘目录与文件的一种抽象表示形式,通过 File 类提供的方法可以实现目录与文件的创建、删除及属性的获取。

1. 构造方法

```
File(String pathname)
File(File parent, String child)
```

其中,参数 pathname 指定路径名;parent 表示上一级目录的 File 对象,字符串 child 表示子目录或文件名称。

2. 获取文件属性

```
String getName()           //返回 File 对象的文件名或者目录名
String getAbsoluteFile()   //返回 File 对象的绝对路径
```

```
long length()                    //获取文件长度
long lastModified()              //获取 File 对象最后一次修改的时间
boolean exists()                 //测试 File 对象对应的文件或目录是否存在
boolean isDirectory()            //测试 File 对象是否是目录
boolean isFile()                 //测试 File 对象是否是文件
boolean isHidden()               //测试 File 对象是否是隐含文件
boolean canRead()                //测试 File 对象是否可读
boolean canWrite()               //测试 File 对象是否可写
```

3. 操作文件及目录

```
boolean createNewFile()          //File 对象对应的文件不存在时,将创建一个新的空文件
boolean delete()                 //删除 File 对象对应的文件
boolean renameTo(File newName)   //重命名当前 File 对象
File[] listFiles()               //返回当前 File 对象对应的目录下的所有文件及子目录
boolean mkdir()                  //创建 File 对象对应的目录。创建成功则返回 true,否则返回 false
```

12.3.2 流的基本概念

计算机应用系统通常涉及多种不同的输入/输出数据源,如键盘、文件、网络、打印机及控制台等。对多种不同数据源的输入/输出操作,其核心需求即稳定点需求,是数据的输入及输出,Java 通过数据流的方式实现了不同数据源的统一操作,大大简化了数据的输入/输出操作。

按照不同的分类方式,数据流可以分为以下几种类型。

1. 输入流和输出流

以当前应用程序为参照读取数据到程序,则称为输入流;反之,从程序输出数据,则称为输出流。如程序一方面从键盘通过输入流获取数据,另一方面可以将程序运行的结果通过输出流送往打印机输出。InputStream 和 Reader 是输入流的抽象父类;OutputStream 和 Writer 是输出流的抽象父类。

2. 字节流和字符流

按照数据流中数据单元的不同,数据流分为字节流与字符流。字节流由 8 位字节组成;字符流操作的数据是 16 位的字符数据。InputStream、OutputStream 是抽象的字节流父类;Reader、Writer 是抽象的字符流父类。

3. 节点流和包装流

按照数据流功能的不同,数据流分为节点流与包装流。应用程序从(向)特定的目标设备读(写)数据的流,称之为节点流;应用程序通过一个间接流对节点流进行重新封装,以期更加高效或者灵活方便地读写各种类型的数据,这里的间接流就称为包装

流或封装流。提供缓存功能的 BufferedReader、BufferedWriter、BufferedInputStream、BufferedOutputStream 及对象流 ObjectInputStream、ObjectOutputStream 都是常见的包装流。

12.3.3 字节流与字符流

InputStream、OutPutStream 是输入字节流及输出字节流的抽象父类；Reader、Writer 是输入字符流及输出字符流的抽象父类。

1. 输入流——InputStream 与 Reader

InputStream 与 Reader 分别是字节、字符输入流的抽象父类，提供如下常用方法。

```
int read()                          //从字节(字符)输入流中读取并返回 1 字节(字符)
int read(byte[]/char[] data)        //从字节(字符)输入流中读取若干字节(字符),并保
                                    //存于数组 data 中,返回读取的字节(字符)数。若遇
                                    //到文件结尾,则返回-1
long skip(long length)              //从字节(字符)输入流中跳过 length 字节(字符),方
                                    //法返回实际跳过的字节(字符)数
```

2. 输出流——OutputStream 与 Writer

OutputStream 与 Writer 分别是字节、字符输出流的抽象父类，提供如下常用方法。

```
void write(int b)                   //输出字节(字符)b 到字节(字符)的输出流
void write(byte[]/char[] data)      //输出数组 data 中若干字节(字符)到字节(字符)的
                                    //输出流
```

3. 字节流子类——文件输入/输出字节流

字节流 InputStream 与 OutputStream 是抽象类，它们无法实例化对象，只可以通过实例化子类对象完成数据流的操作。FileInputStream 与 FileOutputStream 是 InputStream 与 OutputStream 的子类。通常用于实现二进制文件的输入/输出操作。可以采用以下几种方式创建文件输入/输出字节流。

```
FileInputStream(File file)
FileInputStream(String filename)
FileOutputStream(File file)
FileOutputStream(File file, boolean append)
```

其中，参数 append 标识是否采用追加模式。

4. 字符流子类——文件输入/输出字符流

与字节流子类相似，Java 提供字符流子类，通过实例化子类对象完成多种字符流的输入/输出操作。FileReader/FileWriter 是 Reader 与 Writer 的子类，用于实现文本文件

的输入/输出操作。构建文件输入/输出字符流的方法为

```
FileReader(File file)
FileReader(String filename)
FileWriter(File file)
FileWriter(File file, boolean append)
```

5. 字节流与字符流的操作流程

字节流或者字符流的操作可以归纳为如下步骤。
(1) 创建数据源或者目标对象。
如创建待访问的文件对象、创建缓冲区数组或者字符串等。
(2) 创建输入/输出流对象。
如果输入/输出的数据是文本内容,则创建字符流子类对象;如果输入/输出的数据是二进制数,则创建字节流子类对象。
(3) 通过输入/输出流对象的读写操作完成数据的输入/输出。
(4) 关闭数据流。
调用数据流的 close()方法关闭数据流,以释放资源。
例 12-1 示范了字节流的操作过程。

例 12-1

```
public class FileOutputStreamTest {
    public static void main(String[] args) throws IOException {
        FileInputStream fis = new FileInputStream("src\test\FileTest.java");
        FileOutputStream fos = new FileOutputStream("src\test\FileTestCopy.
                                                    java");
        byte[] buffer = new byte[1024];        //定义缓存数据块
        while(fis.read(buffer)>0) {
            fos.write(buffer);
        }
        fos.close();
        fis.close();
    }
}
```

运行例 12-1,以字节流的方式实现文件的复制。对于文本文件,也可以以字符流的方式进行读写操作。

12.3.4 包装流

通过对节点流进行重新封装可以产生包装流。通常,包装流相对于节点流可以提供更高效或者更方便的输入输出方式。

1. 缓冲流

在通过缓冲流进行读取或写入时对数据进行缓存,以减少输入/输出的次数,提高输入/输出的效率。Java 提供 4 种缓冲流:BufferedInputStream、BufferedOutputStream、BufferedReader 与 BufferedWriter。常用的构造方法为

```
BufferedInputStream(InputStream in)
BufferedInputStream(InputStream in, int size)
```

封装输入字节流 in,构建缓冲输入字节流,参数 size 指定缓冲区的大小。

```
BufferedOutputStream(OutputStream out)
BufferedOutputStream(OutputStream out, int size)
```

封装输出字节流 out,构建缓冲输出字节流,参数 size 指定缓冲区的大小。

```
BufferedReader(Reader in)
BufferedReader(Reader in, int sz)
```

封装输入字符流 in,构建缓冲输入字符流,参数 size 指定缓冲区的大小。

```
BufferedWriter(Writer out)
BufferedWriter(Writer out, int sz)
```

封装输出字符流 out,构建缓冲输出字符流,参数 size 指定缓冲区的大小。

2. 转换流

通过转换流将字节流转换为字符流。Java 提供两种转换流:InputStreamReader 与 OutputStreamWriter。常用的构造方法为

```
InputStreamReader(InputStream in, Charset cs)
InputStreamReader(InputStream in)
```

将输入字节流 in 转换为输入字符流。InputStreamReader 转换流从输入字节流读取字节,按照指定的字符集 cs 将字节解码为字符。

```
OutputStreamWriter(OutputStream out, Charset cs)
OutputStreamWriter(OutputStream out)
```

将输出字节流 out 转换为输出字符流,参数 cs 指定编码方式。

3. 包装流的操作流程

包装流的操作步骤为:
(1) 定义节点流对象;
(2) 构造包装流对象,实现对节点流对象的封装;
(3) 通过包装流对象的读写完成数据的输入/输出操作;
(4) 关闭包装流对象。

12.3.5 对象序列化及对象流

1. 对象序列化与反序列化

把内存中的对象数据转换为字节序列,以便对象能够保存到磁盘或者通过网络传递到其他网络节点,这个过程称为序列化。

与序列化相反,把磁盘文件或者通过网络传输得到的字节序列恢复成 Java 对象的过程称为反序列化。

2. 通过对象流实现序列化

(1) 序列化接口。

Java 规定只有实现了序列化接口 Java.io.Serializable 或者 Java.io.Externalizable 的类,其对象才可以被序列化。实现 Serializable 接口的类无须重写任何方法,系统提供默认的序列化方式;实现 Externalizable 接口的类需要重写以下两个方法,自定义序列化及反序列化的行为。

```
public void writeExternal(ObjectOutput out)
public void readExternal(ObjectInput in)
```

注意:

① 实现 Serializable 接口的类必须确保类的成员变量都是可以序列化的:Java 基本数据类型、String 类型或者也实现了 Serializable 接口的引用类型,否则在进行序列化操作时会抛出 java.io.NotSerializableException 异常。

② 如果对象中存在一些不想被序列化的数据,可以使用关键字 transient 进行修饰。

③ 实现 Serializable 接口的类,其父类(包括间接父类)应该是可序列化或者拥有无参的构造方法,否则执行反序列化时会抛出 java.io.InvalidClassException 异常。如果父类没有实现序列化接口,仅定义了无参的构造方法,则在序列化时,该父类的成员变量无法保存到字节流中;在反序列化时,将得到 0 或者 null。

(2) 对象输入/输出流。

Java 提供了 ObjectInputStream 与 ObjectOutputStream 类读取与写入对象信息,称为对象输入/输出流。类的构造方法定义为

```
ObjectInputStream (InputStream in)
ObjectOutputStream (OutputStream out)
```

封装输入字节流 in/输出字节流 out,定义对象输入/输出流。ObjectInputStream 与 ObjectOutputStream 类提供如下方法输入与输出可序列化对象。

```
Object readObject()
void writeObject(Object obj)
```

(3) 对象的序列化与反序列化。

对象的序列化可以通过以下 4 步实现：

① 创建节点流，定义输出源和输入源；

② 实例化 ObjectOutputStream 和 ObjectInputStream 类的对象，封装节点流；

③ 通过对象输出/输入流提供的 writeObject() 和 readObject() 方法实现对象的序列化与反序列化；

④ 关闭输出/输入流。

注意：如果向节点流序列化多个对象，则在反序列化时需要按照写入的顺序进行恢复。

例 12-2 示范了对象的序列化与反序列化。

例 12-2

```
class Organization implements Serializable{                    ①
    private String name;
    private String addr;
    public Organization(String name, String addr) {
        this.name = name;
        this.addr = addr;
    }
    public Organization() {
    }
    @Override
    public String toString() {
        return "Organization [name=" + name + ", addr=" + addr + "]";
    }
}
```

上述代码自定义了 Organization 类，描述了组织名称、所在地等组织信息。

```
public class User implements Serializable{
    private int uid;
    private String uname;
    private String upwd;
    private Organization workFor;
    public String toString() {
        return uid+" "+uname+" "+upwd+" "+ workFor;
    }
    public User() {
    }
    public User(int uid,String uname,String upwd,Organization workFor) {
        this.uid = uid;
        this.uname = uname;
        this.upwd = upwd;
        this.workFor = workFor;
    }
}
```

定义可序列化类 User，要保证类的成员变量都是可以序列化的，如代码行①定义类 Organization 可序列化。

```java
public class ReadObjectTest {
    public static void main(String[] args) throws IOException,
                                                ClassNotFoundException {
        ObjectOutputStream bos = new ObjectOutputStream(
                                new FileOutputStream("user.dat"));
        bos.writeObject(new User(1,"李白","1234",
                                new Organization("唐朝","长安")));
        bos.writeObject(new User(2,"杜甫","5678",
                                new Organization("唐朝","洛阳")));
        bos.close();
        ObjectInputStream bis = new ObjectInputStream(
                                new FileInputStream("user.dat"));
        while(true) {
            try {
                System.out.println((User)bis.readObject());
            }catch(EOFException e) {
                break;
            }
        }
        bis.close();
    }
}
```

12.4 个人通讯录（七）设计步骤

12.4.1 数据库设计

修改数据库 book 中联系人表 contracts 的设计，设计界面如图 12-1 所示。

增加 image 字段，类型为二进制大对象（binary large object，blob）。在数据库系统中，blob 是常用来存储二进制文件的字段类型。

12.4.2 数据实体类 Contract

修改已有的 Contract 类，增加成员变量 byteImg，缓存联系人头像数据。具体代码如下所示，省略与个人通讯录（六）相同的代码。

12.4.2

```java
public class Contract {
    private byte[] byteImg;
    public Contract(int id, String name, String gender, String email, Set<String>
```

```
    phones, byte[] image) throws Exception {
        this.id = id;
        setName(name);
        setGender(gender);
        setEmail(email);
        setPhones(phones);
        setImage(image);
    }
}
```

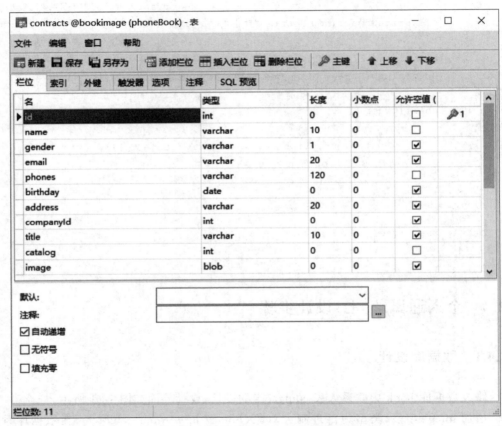

图 12-1　联系人表 contracts 设计视图

12.4.3　数据访问类 ContractDaoImpl

dao 层需要重新定义 ContractDaoImpl 类，以增加联系人关于头像信息的处理。

1. 查询联系人

查找并返回满足条件的联系人列表，同时缓存联系人头像，方法定义如下。

```java
public List<Contract> searchByName(String name) throws Exception {
    ResultSet rs = pst.executeQuery();
    while (rs.next()) {
        //省略与个人通讯录(六)相同的代码
        Blob image = rs.getBlob(11);
        byte[] buf = null;
        if (image != null) {
            buf = image.getBinaryStream().readAllBytes();        //缓存头像数据
        }
    }//while 语句结束
}
```

2. 添加和修改联系人

添加和修改联系人操作都需要实现将包含头像的联系人信息上传至数据库。在个人通讯录(六)的基础上,与头像上传相关的代码定义如下。

```java
public boolean add(Contract c) throws Exception {
    PreparedStatement state = conn.prepareStatement(sql);
    InputStream is = null;
    if (c.getImage() != null && c.getImage().length!= 0) {
        is = new ByteArrayInputStream(c.getImage());
        state.setBlob(10, is);
    } else {
            state.setNull(10, Types.BLOB);
    }
}
```

12.4.4 图形用户界面 ContractDetailGUI

增加头像信息的"编辑(添加)联系人"对话框如图 12-2 所示。

用户单击"单击上传头像"链接会弹出选择文件对话框,选择头像文件后单击"确认"按钮后读取并显示头像。

在个人通讯录(六)的基础上,与头像选择和显示相关的代码定义如下。

12.4.4

```java
public class ContractDetailGUI extends JDialog {
    private final static int MAX_BLOB_LENGTH = 65 * 1024;    //描述 MySQL blob 类型
                                                             //的存储限制
    private JLabel lblImage;                                 //显示头像的标签组件
    private byte[] byteImg;                                  //头像缓存数组
    lblImage.addMouseListener(new MouseAdapter() {           //单击选择头像
        @Override
        public void mouseClicked(MouseEvent e) {
            JFileChooser chooser = new JFileChooser(new
```

图 12-2 增加头像信息的"编辑(添加)联系人"对话框

```
    File(ContractDetailGUI.class.getResource("/").getFile()));
FileNameExtensionFilter filter = new FileNameExtensionFilter("图
    像文件", "jpg", "png");                    //定义文件过滤器
chooser.setFileFilter(filter);
if (JFileChooser.APPROVE_OPTION = = chooser.showOpenDialog(
                                ContractDetailGUI.this)) {
    File selectedFile = chooser.getSelectedFile(); //获得选中的文件
                                                    //对象
    if (selectedFile.length() > MAX_BLOB_LENGTH) {
        JOptionPane.showMessageDialog(ContractDetailGUI.this,
            "上传的图像文件过大,超出范围");
        return;
    }
```

```
            FileInputStream fis;
            try {
                fis = new FileInputStream(selectedFile);
                byteImg = fis.readAllBytes();
                fis.close();
            } catch (IOException e1) {
                JOptionPane.showMessageDialog(ContractDetailGUI.this,
                                                            "读文件出错");
            }
            displayImg();           //显示头像
        }//if 语句结束
    }//mouseClicked 方法结束
});
private void displayImg() {
    if (byteImg != null && byteImg.length > 0) {
        lblImage.setIcon(new ImageIcon(byteImg));
        lblImage.setText("");
    }
}
```

12.5 练一练

1. 编程实现对选定的磁盘文件按照异或算法进行加密、解密处理。

程序运行效果如图 12-3 所示。

提示：使文件中的每个字节与给定密钥进行异或操作即可完成加密或解密处理，如 cipyerText[i]＝cipyerText[i]^secretKey[i％secretKey.length]。

图 12-3　程序运行效果图

2. 编程实现，显示指定路径下的所有文件及子目录(包含子目录下的所有文件)。

第 13 章　多线程——哲学家就餐问题

13.1　哲学家就餐问题介绍

哲学家就餐问题是并行计算中解决多线程同步的经典问题：假设有 5 位哲学家围坐在一张圆形餐桌旁，每两人中间有一支筷子，餐桌中间有一大碗米饭，如图 13-1 所示；每位哲学家交替进行思考与就餐，哲学家思考的时候停止就餐；吃东西时停止思考，并且只有当哲学家拿到其左右手边的两支筷子时才能就餐。设计就餐策略，使得哲学家在就餐时不会因为拿筷子而出现死锁问题。

图 13-1　哲学家就餐问题

13.2　程序设计思路

分析哲学家就餐问题可知，5 位哲学家分别对应 5 个并发线程，并且相邻的两个哲学家（线程）互斥地访问他们之间的筷子（共享资源）。程序设计的目标是避免进程之间发生死锁，所以程序设计的基本思路是从理解进程的死锁含义入手分析死锁产生的原因，尤其是死锁产生的必要条件，从而最大限度地避免、预防及解除线程的死锁。

13.3　关键技术

13.3.1　多线程的基本概念

Java 语言对用户进行多线程程序设计提供语言级的支持。应用程序通过引入多线程技术，可充分利用系统资源，提高应用程序的执行效率。多线程程序设计涉及以下几个基本概念。

1. 程序

程序是一个静态概念，指计算机能够识别与执行的指令集合及指令执行时所需要的数据集合，如微信、Office 办公软件等都是应用程序。

2. 进程

进程是一个动态概念，指应用程序的一次运行，包括从代码加载、分配地址空间到执行指令序列的一个完整过程。进程拥有自己的生命周期与独立的地址空间。多个进程都独立地运行在 CPU 上，互不干扰。通常，一个进程也不能直接访问其他进程拥有的地址空间。

3. 线程

线程也称轻量级的进程，是进程中可以独立执行的子序列。一个线程代表进程中一个单一顺序的执行流。一个进程可以同时并发多个线程，每个线程并行执行不同的任务。每个线程可以拥有私有的堆栈，定义仅有当前线程可以访问的局部变量，但是线程不独享系统资源，进程的所有线程共享系统资源。线程可以独立完成任务，也可以互相通信，协同完成进程所要完成的任务。

13.3.2 实现多线程的方法

每个进程至少拥有一个线程，称为主线程。应用程序中，除主线程之外，还可以创建多个线程，Java 语言提供了以下 3 种实现线程的方式。

1. 继承 Thread 类创建线程子类

创建 java.lang.Thread 子类及使用多线程的流程如下：
（1）自定义线程类，重写 run() 方法，实现线程的功能；
（2）实例化线程对象；
（3）调用 start() 方法，启动线程。
例 13-1 示范了继承 Thread 类创建并启动多线程的方法。
例 13-1

```java
public class ThreadTest extends Thread{
    public static void main(String[] args) {
        System.out.println("启动主线程");
        Thread sub1 = new SubThread("线程 1",8);
        Thread sub2 = new SubThread("线程 2",10);
        sub1.start();
        sub2.start();
        System.out.println("主线程结束");
    }
```

```java
}
public class SubThread extends Thread{
    private int count;                    //定义输出序列的初始值
    public SubThread(String name,int count) {
        super(name);
        this.count = count;
    }
    public SubThread(String name) {
        this(name,0);
    }
    public void run() {                   //重写父类 Thread 的线程体
        System.out.println(getName()+"开始: ");    //①
        while(count < 15) {
            count++;
            System.out.println(getName()+"----->"+ getCount());
        }
        System.out.println(getName()+"结束。");
    }
}
```

例 13-1 定义了 Thread 类的子类 SubThread，重写了 run()方法，定义线程执行体。在 run()方法中，代码行①使用 Thread 类提供的方法 getName()获取线程名称。类 ThreadTest 中定义了 main()方法，显式创建并启动两个线程 sub1 与 sub2。实际上，当前应用程序先后启动了 3 个线程：main 方法所在的主线程、sub1 线程、sub2 线程。

例 13-1 的运行结果如图 13-2 所示。

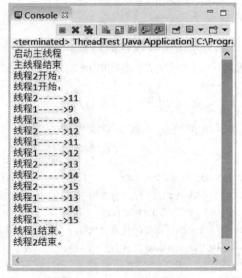

图 13-2 例 13-1 的运行结果

程序首先启动主线程,创建并启动两个线程 sub1 与 sub2,然后主线程结束。这时系统仅剩两个工作线程 sub2 与 sub1。线程调度器首先调度并执行 sub2 线程,输出"线程 2 开始:",接着 sub2 线程的执行时间(也称一个 CPU 时间片)结束,sub2 线程让出 CPU;操作系统再次调度并执行 sub1 线程,输出"线程 1 开始:"。以此往复,两个线程交替执行。由于线程被调度的不确定性,所以每次执行例 13-1 都可能会得到不同的输出结果。

2. 实现 Runnable 接口创建线程子类

实现 java.lang.Runnable 接口创建线程子类及使用多线程的流程如下。

（1）自定义线程类,实现 Runnable 接口。Runnable 接口仅包含一个方法（public void run()）,定义线程执行体,实现类需要重写该方法,以实现线程的功能。

（2）实例化自定义线程类对象。

（3）以步骤（2）中构建的线程对象为参数,调用 Thread 类的构造方法实例化 Thread 对象。

（4）调用 Tread 对象的 start()方法启动线程。

例 13-2 示范了实现 Runnable 接口创建并启动多线程的方法。

例 13-2

```java
public class RunnablePro implements Runnable {
    private int count;
    public RunnablePro() {
    }
    public RunnablePro(int count) {
        this.count = count;
    }
    public int getCount() {
        return count;
    }
    public void run() {     //重写 Runnable 接口中的 run()方法
        System.out.println(Thread.currentThread().getName()+ "开始：");//②
        while (count < 15) {
            count++;
            System.out.println(Thread.currentThread().getName()
                                    +"-----> " + getCount());
        }
        System.out.println(Thread.currentThread().getName() + "结束。");
    }
}
public class RunnableTest {
    public static void main(String[] args) {
        System.out.println("启动主线程");
        RunnablePro pro1 = new RunnablePro(8);
        RunnablePro pro2 = new RunnablePro(10);
```

```
            new Thread(pro1, "线程 1").start();
            new Thread(pro2, "线程 2").start();
            System.out.println("主线程结束");
        }
    }
```

例 13-2 定义了 Runnable 接口的实现类 RunnablePro,重写了 run()方法,定义线程执行体。实现了与例 13-1 相同的功能,RunnablePro 类中需要获取线程的名称。例 13-1 的代码行①使用 Thread 类提供的方法 getName()获取线程名称,但是此处不可以直接使用 Thread 类提供的方法。程序首先通过 Thread.currentThread()获取当前正在执行的线程 Thread 对象,然后才能调用 getName()获取线程名称,如代码行②所示。

3. 实现 Callable 接口创建具有返回值的线程子类

Java 5 开始提供了 Callable 接口。实现 Callable 接口创建线程子类及使用多线程的流程如下。

(1) 自定义线程类,实现 Callable 接口,重写 call()方法,实现线程的功能。

java.util.concurrent.Callable 接口与 Runnable 接口相似,两者的不同之处在于,Callable 接口中定义的 call()方法可以有返回值,允许抛出异常。

Callable 接口的定义为

```
Interface Callable<V>{
    V call() throws Exception;
}
```

其中,V 定义 call()方法的返回值类型。call()方法是线程执行体,实现类需要重写该方法。

(2) 实例化自定义线程类对象。

(3) 以步骤(2)中构建的线程对象为参数构建 FutureTask<V>对象。

java.util.concurrent.FutureTask<V>类实现 Future<V>接口和 Runnable 接口,封装了 Callable<V>对象。FutureTask<V>类提供如下构造方法。

```
FutureTask(Callable<V> callable)
```

(4) 以步骤(3)中创建的 FutureTask<V>对象调用 Thread 类的构造方法实例化 Thread 对象。

(5) 调用 Tread 对象的 start()方法,启动线程。

(6) 调用 FutureTask 对象的 get()方法获得子线程的执行结果。

```
V get()
```

获取 Callable<V>对象执行 call()方法的返回值。调用该方法将导致主线程阻塞,直到 call()方法执行结束并返回为止。

```
V get(long timeout,TimeUnit unit)
```

调用该方法将导致主线程最多阻塞参数指定的时间。其中，参数 timeout 定义最大等待时间，unit 指定时间单位。如果在指定时间后 call() 方法还没有结束，则抛出 TimeoutException 异常。

例 13-3 示范了实现 Callable<V>接口创建并启动多线程的方法。

例 13-3

```java
public class CallablePro implements Callable< Integer> {            //①
    private int count;
    public CallablePro(int count) {
        this.count = count;
    }
    @Override
    public Integer call() throws Exception {
        System.out.println(Thread.currentThread().getName() + "开始：");
        while (count < 15) {
            count++;
            System.out.println(Thread.currentThread().getName()
                                +"----> " + getCount());
        }
        System.out.println(Thread.currentThread().getName() + "结束。");
        return count;
    }
}
public class CallableTest {
    public static void main(String[] args) throws Exception{
        System.out.println("启动主线程");
        CallablePro pro1 = new CallablePro(8);                      //②
        CallablePro pro2 = new CallablePro(10);
        FutureTask< Integer> f1 = new FutureTask< Integer> (pro1);   //③
        FutureTask< Integer> f2 = new FutureTask< Integer> (pro2);
        new Thread(f1, "线程 1").start();                            //④
        new Thread(f2, "线程 2").start();
        System.out.println(f1.get());                                //⑤
        System.out.println(f2.get());
        System.out.println("主线程结束");
    }
}
```

例 13-3 首先定义了实现 Callable 接口的类 CallablePro，如代码行①所示；重写了 call() 方法，call() 方法的功能与例 13-2 的 run() 方法的功能相似，唯一的不同之处在于，call() 方法中增加了 return count; 语句。

在主程序 CallableTest 中实例化 CallablePro 对象 pro1 与 pro2，如代码行②所示；代码行③封装 CallablePro 对象，构建 FutureTask 对象；代码行④以 FutureTask 对象为参

数,实例化并启动 Thread 对象;主线程调用了 FutureTask 对象的 get()方法,获取子线程的执行结果,即 call()方法的返回值,如代码行⑤所示,最后主线程向控制台输出"主线程结束"。

例 13-3 的运行结果如图 13-3 所示。

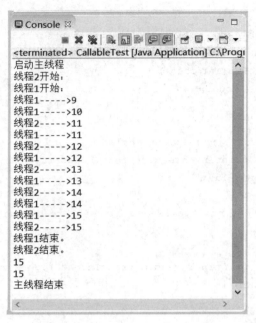

图 13-3　例 13-3 的运行结果

分析例 13-3 的运行结果可以得出本程序的运行过程:系统首先调度主线程,控制台输出"启动主线程";然后主线程分别启动"线程 1"与"线程 2",此时,系统中存在 3 个工作线程,分别是主线程、"线程 1"与"线程 2"。主线程执行代码行⑤时发生阻塞,直到子线程的 call()方法执行结束才能获得返回值。接下来线程调度器多次调度"线程 2"与"线程 1",分别执行各自的线程体 call()方法;"线程 1"与"线程 2"的 call()方法执行结束,主线程通过 get()方法获取线程体的执行结果,控制台输出"15",最终主线程输出"主线程结束"。

13.3.3　线程的生命周期

线程从创建到死亡的完整过程称为线程的生命周期。线程在生命周期中通常会经历 5 个状态,分别为新建、就绪、运行、阻塞与死亡。线程的生命周期如图 13-4 所示。

1. 新建状态

实例化线程对象,初始化对象中的成员变量,线程处于新建状态。

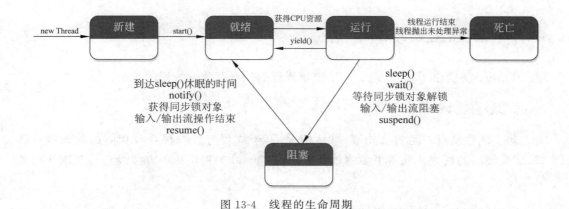

图 13-4　线程的生命周期

2. 就绪状态

调用线程对象的 start() 方法，线程处于就绪状态，等待 JVM 线程调度器的调度。

注意：仅当线程处于新建状态时，才能调用线程对象的 start() 方法，否则当线程已经执行完线程体之后，再次调用 start() 方法会抛出 IllegalThreadStateException 异常。

3. 运行状态

就绪状态的线程在获得 CPU 控制权之后开始执行线程的 run() 方法或者 call() 方法，线程进入运行状态。

4. 阻塞状态

（1）运行状态的线程发生如下情况会进入阻塞状态。

① 调用线程 sleep() 方法。
② 调用线程 wait() 方法，等待通知。
③ 线程等待其他线程释放同步锁对象。
④ 等待输入/输出操作结束。
⑤ 线程调用 suspend() 挂起方法。

（2）阻塞状态的线程发生如下情况时会恢复就绪状态。

① 休眠线程到达休眠时间后。
② 调用 wait() 的线程，获得通知 notify() 或者 notifyAll()。
③ 获取同步锁对象。
④ 输入/输出操作结束。
⑤ 线程调用 resume() 恢复方法。

5. 死亡状态

如果运行状态的线程运行结束或者线程抛出未处理异常，则线程结束，进入死亡状态。

13.3.4 线程的控制方法

Thread 类提供了很多方法,用于控制线程的执行及改变线程的状态。

1. 线程的优先级

每个线程具有一定的优先级,默认情况下,子线程与父线程具有相同的优先级。通常,优先级高的线程将获得更多的调度机会。Thread 类提供如下方法设置与获取线程的优先级:

```
public final void setPriority(int newPriority)    //设置当前线程的优先级
public final int getPriority()                    //获取当前线程的优先级
```

注意:参数 newPriority 的取值范围为 1~10。
Thread 类中定义了如下静态常量以表达优先级的大小。

```
public static final int MAX_PRIORITY       //最高优先级,值 10
public static final int NORM_PRIORITY      //默认优先级,值 5
public static final int MIN_PRIORITY       //最低优先级,值 1
```

2. sleep()方法

```
public static void sleep(long millis)
```

暂停当前线程的执行,使线程进入阻塞状态,休眠 millis 毫秒。

3. yield()方法

```
public static void yield()
```

当前进程出让 CPU 控制权,由运行状态进入就绪状态,等待 JVM 的重新调度。

注意:当前线程让出 CPU 控制权之后,只有优先级大于或等于当前线程的就绪线程才可能被调度,进入运行状态。

4. 守护线程

线程可以分为用户线程与守护线程两类,两者几乎没有区别,唯一的不同之处在于当所有的用户线程运行结束时,会"杀死"守护线程,结束应用程序的执行;反之,只要存在运行中的用户线程,程序就不会终止。

使用守护线程时需要注意以下几点:

(1) 调用 Thread 对象的 setDaemon(true)方法可以将用户线程转换为守护线程。

(2) 调用 setDaemon(true)方法设置守护线程必须在启动线程之前,否则会抛出 IllegalThreadStateException 异常。

(3) 守护线程中创建的子线程也是守护线程。

13.3.5　线程的同步

多线程程序中同时存在多个并发线程。一方面,多个线程独立、异步执行;另一方面,多个线程可以共享数据。因此,必须采用同步机制控制多个线程对共享资源的访问,否则无法保证程序运行结果的正确性。

例 13-4 示范了多个线程访问共享数据时出现的问题。

例 13-4　定义共享资源 ShareRes 类。

```java
public class ShareRes {
    private int count = 10;                //初始资源数目
    public int getCount() {                //获取当前可用资源的编号
        return count;
    }
    public void consumeCount() {           //消耗资源
        if (count > 0) {
            System.out.println(Thread.currentThread().getName()
                                        +"消耗资源："+getCount());
            count--;
        }
    }
}
```

定义消耗共享资源的线程类 SharePro。

```java
public class SharePro implements Runnable{
    private ShareRes res;
    public SharePro(ShareRes res) {
        this.res = res;
    }
    public void run() {
        while(true) {
            res.consumeCount();     //消耗资源
            try {
                Thread.sleep(100);
            } catch (InterruptedException e) {
                e.printStackTrace();
            }
        }
    }
}
```

主程序定义如下。

```java
public class SynchronizedTest {
```

```
        public static void main(String[] args) {
            ShareRes res = new ShareRes();
            SharePro p1 = new SharePro(res);
            SharePro p2 = new SharePro(res);
            new Thread(p1,"线程 1").start();
            new Thread(p2,"线程 2").start();
        }
    }
```

主程序中定义了共享资源 res，分别声明两个消耗共享资源的对象 p1 与 p2，以 p1 与 p2 为目标对象创建并启动"线程 1"与"线程 2"。运行例 13-4，得到如图 13-5 所示的运行结果。

分析运行结果，发现"线程 2"首先被调度，调用 res.consumeCount()方法输出 10 号资源被消耗，在执行"count――;"之前，"线程 2"出让 CPU，进入就绪状态；"线程 1"被调度，同样调用 res.consumeCount()方法，此时 count 变量的值依然是 10，所以 10 号资源被消耗了两次，程序出现了错误的结果。

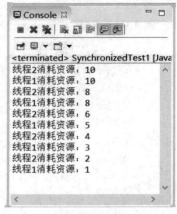

图 13-5　例 13-4 的运行结果

1. 同步代码块及同步方法

针对上述问题，Java 引入了同步代码块与同步方法。同步代码块的定义为

```
synchronize(obj){
    …//同步代码块
}
```

Java 称同步代码块中的 obj 对象为同步监视器。当一个线程访问某个对象的同步代码块时，首先对同步监视器对象上锁。在加锁期间，其他任何线程都无法再访问该对象上的任何同步代码，直到访问线程执行结束（或者抛出未处理的异常），才会释放对该对象的锁定。

与同步代码块相似，应用程序中可以定义同步方法，在方法的定义之前增加 synchronized 即可：

```
[访问修饰符] synchronized 返回值类型 方法名(参数){
    …
}
```

同步方法的同步监视器对象是 this 对象。

2. 同步锁

Java 5 开始提供功能更强大、使用方式更灵活的同步控制机制——同步锁。Lock 接口及其实现类 ReentrantLock（可重入锁）以显式定义同步锁对象的方式控制多个线程对共享资源的独占访问。

通常，使用同步锁功能的代码模式定义为

```java
class SharedResource{
    private ReentrantLock lock = new ReentrantLock();      //定义同步锁
    public void method(){                                   //线程安全的方法
        lock.lock();                                        //加锁
        try{
            ...                                             //线程安全的代码块
        }finally{
            lock.unlock();                                  //解锁
        }
    }
}
```

注意：

（1）一般将解锁操作置于 finally 字句中，这样即使 try 代码块有异常抛出，也会执行解锁操作。

（2）try 代码块就是加锁至解锁之间的代码，它不宜是太耗时间的代码，否则会大大降低程序的执行效率。

使用同步锁机制将例 13-4 中的共享资源 ShareRes 类定义为

```java
public class ShareRes {
    private final ReentrantLock lock= new ReentrantLock();      //①
    private int count = 10;
    public int getCount() {
        return count;
    }
    public void consumeCount() {
        lock.lock();            //加锁
        try{                    //线程安全的代码块
            if (count > 0) {
                System.out.println(Thread.currentThread().getName()
                                    +"消耗资源："+getCount());
                count--;
            }
        }finally {
            lock.unlock();   //释放锁
        }
    }
}
```

上述程序的代码行①定义了同步锁对象，使用 final 关键字保证一个共享资源仅对应一个锁对象。consumeCount()方法使用了与同步方法相似的执行逻辑：首先锁对象加锁，然后执行操作共享资源的代码块，最后释放锁。

3. 死锁

死锁是指多个线程并发执行时因争夺资源而造成的一种互相等待的现象。一旦发生死锁，系统中所有的线程都将处于阻塞状态。多线程程序中，尤其是当系统中出现多个共享资源（同步监视器或者同步锁）时，经常会发生死锁现象。例如两个线程线程 1 与线程 2，如果线程 1 已经拥有对象 A 的锁，但是又需要对对象 B 加锁；同时，线程 2 已经保持对象 B 的锁，又需要对对象 A 加锁，这种情况下这两个线程会互相等待对方释放锁，就会形成死锁。

13.3.6 线程通信

多线程编程不仅需要考虑多个线程同步、互斥地访问共享资源，还需要考虑线程之间的通信问题。Java 提供了如下两种方式实现线程之间的通信。

1. wait()、notify() 及 notifyAll() 方法

wait()、notify() 及 notifyAll() 方法由 Object 类提供。

(1) wait() 方法。

```
public final void wait()
public final void wait(long timeout)
public final void wait(long timeout, int nanos)
```

阻塞当前线程，同时释放对同步监视器的锁定，直到其他线程调用该同步监视器的 notify() 方法或者 notifyAll() 方法唤醒当前线程。若指定时间参数 timeout（毫秒）及 nanos（毫微秒），则当线程阻塞超过指定时间后会自动被唤醒，进入就绪状态。

(2) notify() 方法。

```
public final void notify()           //唤醒单个线程
public final void notifyAll()        //唤醒所有线程
```

注意：wait()、notify() 及 notifyAll() 方法只能在同步代码块或者同步方法中使用，否则会抛出 IllegalMonitorStateException 异常。

2. Condition 对象

对于使用 Lock 接口及其实现类 ReentrantLock（可重入锁）实现同步控制的情况，可以使用 Condition 对象控制多个线程之间的通信。Condition 接口中定义了如下方法。

(1) await() 方法。

```
public void await()
```

阻塞当前线程，将当前线程加入等待队列，同时释放锁对象，直到其他线程调用该

Condition 对象的 signal 方法()或者 signalAll()方法唤醒该线程。

（2）signal()方法。

```
public void signal()                    //唤醒单个线程
public void signalAll()                 //唤醒所有线程
```

注意：调用以上方法的前提是当前线程已经获得与该 Condition 对象绑定的锁对象，否则调用上述方法时会抛出 IllegalMonitorStateException 异常。

使用同步锁功能控制线程同步，Condition 对象控制线程通信的代码模式为

```
class SharedResource{
    private ReentrantLock lock=new ReentrantLock();      //定义同步锁
    //绑定 Condition 对象与锁对象
    private Condition condition=lock.newCondition();
    public void method(){
        lock.lock();                    //加锁
        ...
        try{
            if(条件)
                condition.await();      //使当前线程加入等待队列,并释放锁对象 lock
            else{
                ...
                condition.signalAll();//唤醒 condition 对象上等待的所有线程
            }
        }finally{
            lock.unlock();              //解锁
        }
    }
}
```

13.4 哲学家就餐问题程序设计步骤

13.4.1 死锁的产生

以筷子为竞争资源设置同步监视器，控制相邻线程（哲学家）的互斥访问，程序如下所示。

1. 类 ChopStick

描述筷子的属性，提供同步方法 takeChopstick（放下筷子）及 putChopsticks（拿起筷子），实现互斥地访问共享资源。

```
public class ChopStick {
    private volatile boolean isAvailable = true;        //筷子是否可用
```

13.4.1

```java
        private int id;                                      //筷子的编号
        public ChopStick(int id) {
            this.id = id;
        }
        public synchronized void takeChopstick() {           //拿起筷子
            while (true) {
                try {
                    if (isAvailable) {
                        setAvailable(false);
                        System.out.println("哲学家" + "拿起筷子" + getId());
                        break;
                    } else {
                        System.out.println("哲学家" + "在等待筷子" + getId());
                        wait();
                    }
                } catch (InterruptedException e) {
                    e.printStackTrace();
                }
            }
        }
        public synchronized void putChopsticks() {           //放下筷子
            setAvailable(true);
            System.out.println("哲学家" + "放下筷子" + this.getId());
            notifyAll();
        }
    }
```

2. 类 BundleOfChopSticks

BundleOfChopSticks 代表一支筷子，用来封装筷子数组，提供为哲学家分配筷子的方法。假设有 5 位哲学家，编号分别为 0～4，餐桌上摆放了 5 支筷子，编号分别为 0～4，0 号哲学家左手边的筷子的编号是 0，右手边的筷子的编号是 1；以此类推，最后的 4 号哲学家左手边的筷子的编号为 4，右手边的筷子的编号为 0。

```java
public class BundleOfChopSticks {
    private ChopStick[] chopsticks;
    public BundleOfChopSticks(ChopStick[] chopsticks) {
        this.chopsticks = chopsticks;
    }
    public BundleOfChopSticks(int size) {
        chopsticks = new ChopStick[size];
        for (int i = 0; i < size; i++) {
            chopsticks[i] = new ChopStick(i);
        }
    }
```

```java
//获取编号为id的哲学家指定筷子(位于哲学家的左手边)
public ChopStick getPreChopStick(int id) {
    return chopsticks[id];
}
//获取编号为id的哲学家指定筷子(位于哲学家的右手边)
public ChopStick getNextChopStick(int id) {
    return chopsticks[(id + 1) % chopsticks.length];
}
}
```

3. 类 Philosopher

线程类 Philosopher 用来封装筷子信息，提供哲学家思考及就餐方法，同时定义线程执行体，模拟哲学家交替进行思考与就餐的过程。

```java
public class Philosopher implements Runnable {
    private int index;                          //哲学家的编号
    private BundleOfChopSticks chopSticks;
    public Philosopher(int index, BundleOfChopSticks chopSticks) {
        this.index = index;
        this.chopSticks = chopSticks;
    }
    public void thinking() {
        System.out.println("哲学家" + index + "在思考");
        try {
            Thread.sleep(500);                  //模拟思考过程
        } catch (InterruptedException e) {
            e.printStackTrace();
        }
    }
    public void eating() {
        System.out.println("哲学家" + index + "在吃饭");
        try {
            Thread.sleep(500);                  //模拟就餐过程
        } catch (InterruptedException e) {
            e.printStackTrace();
        }
    }
    @Override
    public void run() {
        while (true) {
            thinking();
            //拿起左手边的筷子
            (chopSticks.getPreChopStick(index)).takeChopstick();
            //拿起右手边的筷子
```

```
            (chopSticks.getNextChopStick(index)).takeChopstick();
            eating();
            //放下左手边的筷子
            (chopSticks.getPreChopStick(index)).putChopsticks();
            //放下右手边的筷子
            (chopSticks.getNextChopStick(index)).putChopsticks();
        }
    }
}
```

4. 类 RoundTableEating

主控程序,定义 5 支筷子,创建并启动 5 位哲学家线程。

```
public class RoundTableEating {
    public static void main(String[] args) {
        Philosopher[] p= new Philosopher[5];
        BundleOfChopSticks chopSticks = new BundleOfChopSticks(p.length);
        for(int index = 0; index < p.length; index++) {
            p[index] = new Philosopher(index,chopSticks);
            new Thread(p[index]).start();
        }
    }
}
```

多次运行程序,会得到如图 13-6 所示的一次运行结果。从运行结果中可以看出,5 位哲学家都处于等待状态:哲学家 4 首先拿起左手边的筷子 4,然后等待其右手边的筷子 0;而筷子 0 已经被 0 号哲学家拿起,0 号哲学家在等待 1 号筷子;1 号筷子已经被 1 号哲学家拿起,1 号哲学家也无法就餐,因为他在等待已经被哲学家 2 拿起来的 1 号筷子……

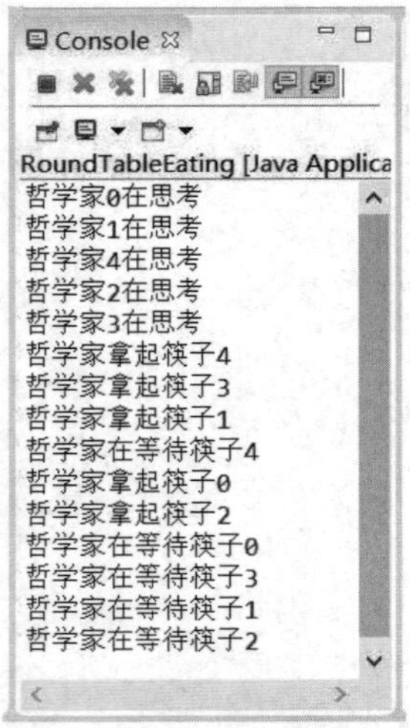

通过分析得到,线程死锁产生的必要条件如下。

(1) 互斥条件:资源是独占且排他使用的,即任意时刻一个资源只能被一个线程占用,其他线程若申请同一个资源,则只能等待直到资源被占有者释放。

(2) 不可剥夺条件:线程获得的资源在未使用完毕之前不能被其他线程强行剥夺,而只能由获得该资源的线程释放。

(3) 请求和保持条件:当一个线程因请求新的资源而阻塞时,继续占有已获得的资源。

(4) 循环等待条件:若干线程之间形成循环等待资源关系。

一方面,只要系统发生死锁,上述 4 个条件必然成立;另一方面,只要上述条件之一不满足,就可以避免死锁的发生。

图 13-6 哲学家就餐问题运行结果(死锁)

哲学家就餐问题可以从破坏"循环等待条件"入手，以避免系统发生死锁，即只有当哲学家可以同时拿起左、右手边的两支筷子时，他才会拿起筷子，否则哲学家只能处于等待状态。

13.4.2 死锁解决方案

13.4.2

使用 Lock 接口及 Condition 对象控制多个线程之间的互斥及通信。

1. 类 BundleOfChopSticks

```java
public class BundleOfChopSticks {
    private boolean[] isOccupied;              //筷子是否被占用
    private final Lock lock = new ReentrantLock();
    private final Condition condition = lock.newCondition();
    public BundleOfChopSticks(int count) {
        isOccupied = new boolean[count];
    }
    //同时拿起左边的 index 及右边的 index+1 的两支筷子
    public void takeChopsticks(int index) {
        lock.lock();
        int nextIndex = ( index + 1) % isOccupied.length;
        try {
            while (isOccupied[index] || isOccupied[nextIndex]) {
                System.out.println("哲学家" + index + "在等待筷子");
                condition.await();
            }
            isOccupied[index] = true;
            isOccupied[nextIndex] = true;
            System.out.println("哲学家" + index + "拿起筷子");
        } catch (Exception e) {
            e.printStackTrace();
        } finally {
            lock.unlock();
        }
    }
    //同时放下编号为 index 及 index+1 的两支筷子
    public void putChopsticks(int index) {
        lock.lock();
        int nextIndex = (index + 1) % isOccupied.length;
        try {
            isOccupied[index] = false;
            isOccupied[nextIndex] = false;
            System.out.println("哲学家" + index + "放下筷子");
```

```
            condition.signalAll();
        } finally {
            lock.unlock();
        }
    }
}
```

2. 类 Philosopher

```
public class Philosopher implements Runnable {
    private int index;
    private BundleOfChopSticks chopSticks;
    //省略构造方法、thinking()方法与eating()方法
    @Override
    public void run() {
        while (true) {
            thinking();
            chopSticks.takeChopsticks(index);     //同时拿起两支筷子
            eating();
            chopSticks.putChopsticks(index);      //同时放下两支筷子
        }
    }
}
```

3. 类 RoundTableEating

主控程序，与前述定义相同：创建并启动 5 位哲学家线程。

13.5 练一练

利用多线程模拟实现医院的预约挂号功能，以保证多位患者可以在不同时间段预约挂号同一位医生。

1. Doctor 类（医生线程）：在医生的预约挂号功能中，医生是共享资源，封装医生姓名与出诊时间，提供挂号功能。

2. Patient 类（患者线程）：封装患者姓名和要预约的医生，在线程体中调用预约医生功能，并打印输出结果。

3. 主程序：初始化若干个医生对象，指定医生的出诊时间。实例化并启动若干患者线程。

第 14 章 网络编程——多线程下载工具

14.1 多线程下载工具功能介绍

网络下载工具可以方便、快捷地从互联网上下载我们需要的文本、图像、音频等各种信息资源。下载时人们最关注的问题毫无疑问是速度；除了速度之外，人们还关注断点续传问题。本章将设计与实现一款基于多线程的、具有断点续传功能的网络下载工具。

系统启动后，用户选择"任务-创建下载任务"菜单选项即可创建新的下载任务，运行效果如图14-1所示。单击"开始"按钮，显示如图14-2所示的资源下载进度。

图 14-1 新建下载任务

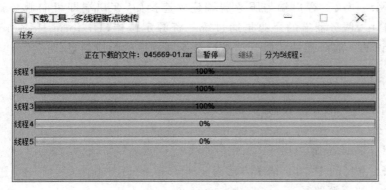

图 14-2 资源下载进度

图 14-2 中的"暂停""继续"按钮模拟了下载的断点续传功能。

14.2 程序设计思路

设计与实现基于多线程的、具有断点续传功能的网络下载工具,需要解决以下问题。

1. 如何下载网络资源

首先是如何定位网络资源,然后才是如何访问和下载资源。

2. 如何提高下载速度

采用多线程技术把待下载的网络资源分割成几个部分同时下载,从而成倍地提高下载速度。

3. 如何做到断点续传

在下载过程中,对于非正常结束的下载线程,记录其断点位置,以备重启系统时读取断点,继续下载。

4. 如何实现多线程并发与 Swing GUI 组件的动态刷新

一方面,启动多个下载线程,分段下载网络资源;另一方面,GUI 需要同步展示各个线程的下载进度。

14.3 关键技术

14.3.1 HTTP

超文本传输协议(Hypertext Transfer Protocol,HTTP)是一种客户端与服务器端请求/响应式的网络协议。客户端与服务器端建立连接后,客户端即可向服务器端发送请求。在 HTTP 中,服务器端的资源是通过统一资源定位器(Uniform Resource Locator,URL)进行标识的。服务器端接收到请求后会返回相应的响应信息。

1. HTTP 请求

HTTP 请求由 3 部分组成,分别是请求行、消息报头、请求正文。
(1) 请求行。
如:GET /userLogin/ HTTP/1.1。
请求方法:常见的 HTTP 请求方法有 get、post、head、delete 等。
URL:请求的资源路径。
HTTP 版本:HTTP 版本号。

(2)消息报头。

消息报头是指从第二行开始到第一个空行结束的部分,一般表示客户端和服务器端进行通信时所需要的一些控制和交互信息。消息报头采用键值对的形式。

(3)请求正文。

当使用 GET 方法向服务器端发送 HTTP 请求时,请求正文为空;当使用 POST 方法时,请求正文封装传递到服务器端的数据。

2. HTTP 响应

服务器端应答客户端的响应信息也由 3 部分组成,分别是状态行、消息报头、响应正文。

(1)状态行。

状态行中包含 HTTP 响应的重要信息:状态码与状态码描述。如常用的状态码 200 表示请求成功;状态码 404 表示服务器端无法找到请求资源。

(2)消息报头。

HTTP 响应消息报头与请求消息报头相似,采用键值对的形式,位于响应消息第二行开始到第一个空行结束的部分。

(3)响应正文。

服务器端返回客户端的 URL 资源,例如浏览器可以识别的 html 文件等。

14.3.2 URL 编程

Java 为网络编程提供了 java.net 包,包内封装了 InetAddress、URL 和 URLConnection 等类,提供多种访问网络资源的方法。

1. InetAddress 类

InetAddress 类代表 IP 地址,该类没有提供构造方法,通过静态方法获取 InetAddress 类的实例对象。

```
static InetAddress getByName(String host)      //根据主机名获取 InetAddress 实例
static InetAddress getByAddress(byte[] addr)   //根据 IP 地址获取 InetAddress 实例
static InetAddress getLocalHost()              //获取与本机对应的 InetAddress 实例
```

2. URL

统一资源定位器(Uniform Resource Locator,URL)用于定位互联网资源,满足格式:

协议名://主机名(IP 地址):端口号/资源名

例如 http://www.tup.tsinghua.edu.cn/upload/books/kj/045669-01.rar。如果省略端口号,则采用默认端口,HTTP 对应 80 端口。

URL 类常用的构造方法定义为

```
URL(String spec)
URL(String protocol,String host, int port, String file)
```

例如:

```
String url = "https://jwc.sdtbu.edu.cn/info/1888/6005.htm";
URLu = new URL(url);
```

获取 URL 实例对象之后,可以通过 openStream()方法获取远程网络资源的输入流对象,通过输入流读取网络资源,实现对远程资源的访问,具体操作过程如下:

(1) 实例化 URL 对象。
(2) URL 对象调用 openStream()方法获取远程 URL 资源的输入字节流。
(3) 读取字节流,对数据进行处理。
(4) 关闭数据流。

3. URLConnection 与 HttpURLConnection

抽象类 URLConnection 代表应用程序和 URL 之间的通信连接。通过 URLConnection 对象可设置参数和请求属性,还可用于读取和写入此 URL 引用的网络资源。HttpURLConnection 类是 URLConnection 的子类,表示与网络资源之间的 HTTP 连接。

通常,创建一个到 URL 的连接、发送请求及读取 URL 网络资源需要以下几个步骤。

(1) 实例化 URL 对象。
(2) URL 对象调用 openConnection()方法,获取与 URL 之间的通信连接 URLConnection 对象。如果通过 HTTP 连接网络资源,则可以将 URLConnection 对象转换为 HttpURLConnection 实例对象。
(3) 设置 URLConnection 对象或者 HttpURLConnection 对象的参数和一般请求属性。

可以采用如下方法设置请求报头字段:

```
void setRequestProperty(String key, String value)     //设置请求报头字段 key 的
                                                      //值为 value
void addRequestProperty(String key, String value)     //为请求报头字段 key 添加
                                                      //值 value
```

(4) 区别 GET 与 POST 请求方式的不同,分别采用不同的方法与远程 URL 对象进行实际连接。

① 采用 GET 发送 HTTP 请求,使用 URLConnection 提供的 connect()方法建立与远程服务器端的实际连接。

② 采用 POST 发送 HTTP 请求,按照如下顺序与远程服务器建立连接。

- 调用 URLConnection 的 setRequestMethod("POST");设置 POST 请求,默认采用 GET 请求方法。

- 调用 URLConnection 的 setDoInput(true);设置该连接可以执行输入操作。
- 调用 URLConnection 的 setDoOutput(true);设置该连接可以执行输出操作。
- 调用 URLConnection 的 getOutputStream()方法获取输出字节流,将来自客户端的请求参数发往服务器端。

（5）访问远程 URL 对象的头字段。

可以使用以下方法访问响应头字段信息。

```
Map<String,List<String>> getHeaderFields()    //获取所有的响应消息头字段
String getHeaderField(String name)            //获取指定响应消息头字段的值
```

（6）通过 URLConnection 对象获取输入流,读取网络资源。
（7）断开 URL 的连接。

14.3.3 基于 TCP 的网络编程

1. TCP 基础

TCP 是传输控制协议（Transmission Control Protocol）的简称。TCP 是面向连接的、可靠的、基于字节流的、端到端的传输层控制协议。当一台计算机与网络上的另外一台计算机通过 TCP 进行通信时,首先会建立一条连接双方的虚拟链路,发送方将数据分割成若干个报文段,依序通过虚拟链路传递给接收方。接收方接收报文段,按照原有的顺序还原报文段。TCP 通过接收方确认和发送方重传机制保证通信的可靠性。

2. 套接字概述

计算机网络上,两个进程通过一个双向的通信连接实现数据的交换,这个双向的通信链路的端点称为套接字（Socket）。Socket 位于 TCP/IP 模型的应用层与传输层之间,Socket 封装了传输层的实现,向上为应用层提供简单的网络通信接口,应用程序开发人员,无须了解 Socket 底层的数据传输机制。

3. 基于 TCP 的 Socket 编程

Java.net 包的类 Socket 和 ServerSocket 分别用于实现基于 TCP 的客户端与服务器端的通信。通常称主动接收其他通信实体连接请求的一方为服务器端,另外一方称为客户端。

（1）服务器端进程。

服务器端的主要任务是创建 ServerSocket 对象,侦听指定端口及 IP 地址,等待连接请求;一旦接收到客户端传来的连接请求,便与客户端建立连接,返回 Socket 对象,开启通信线程以实现与该客户端的通信;主线程继续侦听端口,等待新的连接请求。

服务器端进程通常由以下 5 个步骤组成。

① 创建 ServerSocket 对象。

ServerSocket 常用的构造方法为

```
ServerSocket(int port)
```

```
ServerSocket(int port,int backlog)
ServerSocket(int port,int backlog, InetAddress bindAddr)
```

创建服务器端 ServerSocket 对象，ServerSocket 对象由一个 IP 地址和一个端口号唯一确定。参数 port 是端口号，bindAddr 表示 IP 地址。若构造方法中省略 IP 地址，则使用本机默认 IP 地址。参数 backlog 定义最大连接队列的长度，如果队列已满，则服务器端拒绝接受新的连接请求。

② 调用 ServerSocket 对象的 accept()方法，等待来自客户端的连接请求。

```
Socket accept();
```

如果接收到客户端的连接请求，则返回 Socket 对象；否则服务器端进程处于阻塞状态。

③ 开启通信线程。

通常，客户机-服务器模式采用多对一的通信方式，即一个服务器面向多个客户端同时提供通信服务，所以服务器端通常采用多线程方式。服务器的主线程负责侦听来自客户端的请求，另外设置多个通信子线程，每个通信子线程负责与一个客户端进行通信。

④ 通过服务器端 Socket 对象与客户端进行通信。

Scoket 类提供如下获取字节流对象的方法。

```
InputStream getInputStream();           //获取字节输入流
OutputStream getOutputStream();         //获取字节输出流
```

通信线程中，通过字节流实现服务器端与客户端之间的数据通信。

⑤ 通信完毕，关闭数据流与 Socket 对象。

（2）使用 Socket 的客户端进程。

客户端的主要任务是向服务器端发起连接请求，服务器端应答请求后，客户端返回 Socket 对象，通过该 Socket 对象与服务器端进行通信。

客户端进程通常由以下 3 个步骤组成。

① 创建 Socket 对象。

Socketr 的常用构造方法为

```
Socket(InetAddress/String host, int port)
Socket(InetAddress/String host, int port, InetAddress localAddr, int localPort)
```

客户端通过 Socket 构造方法向服务器端发起连接请求。服务器端应答请求后，客户端返回 Socket 对象。其中，参数 host 表示远程服务器端的 IP 地址，port 表示远程端口，localAddr 与 localPort 用于标识本地的 IP 地址与端口，通常适用于客户端节点具有多个 IP 地址的情形。

② 通过客户端 Socket 对象与服务器端进行通信。

与服务器端类似，通过 Socket 对象获取字节流对象，与服务器端进行通信。

③ 通信完毕，关闭数据流与 Socket 对象。

例 14-1 示范了基于 TCP 的网络 Socket 通信。

服务器端定义类 Server，负责创建 ServerSocket 对象，侦听来自客户端的连接请求。客户端请求到达，则开启通信线程。

例 14-1

```java
public class Server {
    public static void main(String[] args) throws IOException {
        ServerSocket server = new ServerSocket(20000);    //1.创建 ServerSocket 对象
        while (true) {
            Socket s = server.accept();                   //2.等待来自客户端的连接请求
            System.out.println(s.getInetAddress() + "已连接");
            new Thread(new CommuThread(s)).start();       //3.开启通信线程
        }
    }
}
public class CommuThread implements Runnable {           //通信线程
    private Socket s;
    public CommuThread(Socket s) {
        this.s = s;
    }
    @Override
    public void run() {
        OutputStream writer;
        try {
            writer = s.getOutputStream();                //4.通过 Socket 对象与客户端进行通信
            writer.write("连接成功".getBytes());
            writer.flush();
        } catch (IOException e2) {
        }
        InputStream reader = null;
        try {
            reader = s.getInputStream();                 //获取字节输入流
            byte[] b = new byte[1024];
            int length = 0;
            while((length = reader.read(b))!= -1)
                System.out.println(new String(b,0,length));
        } catch (SocketException ee) {                   //客户端退出
            try {
                reader.close();                          //5.关闭数据流与 Socket 对象
                s.close();
            } catch (IOException e) {
                e.printStackTrace();
            }
        } catch (IOException e1) {
            e1.printStackTrace();
```

        }
    }
}

客户端定义 Client 类,负责创建 Socket 对象,向服务器端发起连接请求。与服务器端建立连接之后,客户端负责将用户通过键盘输入的信息发送到服务器端。

客户端还定义了一个接收线程类 ClientThread,用于接收并显示从服务器端发来的信息。客户端的代码如下所示。

```java
public class Client {
    public static void main(String[] args) throws IOException {
        Socket s = new Socket("127.0.0.1",20000);  //1.创建 Socket 对象
        new Thread(new ClientThread(s)).start();  //启动接收线程
        BufferedReader reader = new BufferedReader(
                                    new InputStreamReader(System.in));
        OutputStream writer = s.getOutputStream();  //2.通过 Socket 与服务器端通信
        String line = null;
        while((line = reader.readLine())!=null) {
            writer.write(line.getBytes());
            writer.flush();
        }
    }
}
public class ClientThread implements Runnable{      //接收线程
    private Socket s;
    public ClientThread(Socket s) {
        this.s = s;
    }
    @Override
    public void run() {
        InputStream reader = null;
        try {
            reader = s.getInputStream();
            byte[] b = new byte[1024];
            int length = 0;
            while((length = reader.read(b))!= -1)
                System.out.println(new String(b,0,length));
        } catch (IOException e1) {
            e1.printStackTrace();
        } finally {
            try {
                reader.close();                     //3.关闭数据流与 Socket 对象
                s.close();
            } catch (IOException e) {
```

```
                e.printStackTrace();
            }
        }
    }
}
```

14.3.4 基于 UDP 的网络编程

1. UDP 基础

UDP 是用户数据报协议(User Datagram Protocol)的简称。UDP 是一种无连接的、简单的、不可靠的传输层控制协议。基于 UDP 的通信双方在通信之前无须建立连接,发送方将数据封装成数据报发送出去,接收方接收并完成数据报的转换。UDP 不提供差错控制,传输过程中无法得知数据报是否安全、完整地到达。UDP 主要用于对可靠性要求不高、对实时性要求较高的网络应用场景,如视频会议等。

2. 基于 UDP 的 Socket 编程

Java.net 包的类 DatagramSocket 与 DatagramPacket 用于实现基于 UDP 的网络通信。基于 UDP 通信的双方通常没有明显的服务器与客户端的区别,通常称主动接收来自其他通信实体数据的一方为服务器端或者接收端,另外一方称为客户端或者发送端。

(1) 服务器端(接收端)进程。

服务器端进程通常由以下 4 个步骤组成。

① 创建 DatagramSocket 对象。

DatagramSocket 对象用于实现数据报的发送与接收。服务器端常用的 DatagramSocket 构造方法为

```
DatagramSocket(int port)
DatagramSocket(int port, InetAddress localAddr)
```

创建 DatagramSocket 对象,绑定到本机指定的 IP 地址或默认 IP 地址,指定端口为 port。

与客户端相区别,服务器端通常需要定义明确的 IP 地址与端口号,以便发送方可以确定数据报的目的地址。

② 创建 DatagramPacket 对象。

DatagramPacket 对象用于封装待发送或者接收的数据报。DatagramPacket 提供的构造方法为

```
DatagramPacket(byte[] buf, int offset, int length)
DatagramPacket(byte[] buf, int length)
```

上述两个构造方法用于构造空的接收数据报对象。其中,数组 buf 用于存放接收的数据报;offset 指定存放数据报的开始位置;length 表示最多接收的字节数。

③ 接收 UDP 数据报。

DatagramSocket 类提供如下方法实现数据报的接收。

```
receive(DatagramPacket packet)
```

执行 receive()方法,接收端阻塞,等待来自客户端发来的数据报。一旦有数据到达,服务器端接收数据并填充 DatagramPacket 对象。通过 DatagramPacket 对象可以获取发送方的 IP 地址、端口、接收到数据的长度等信息。

④ 关闭 DatagramSocket 对象。

通信结束,执行 close()方法关闭 DatagramSocket 对象。

(2) 客户端(发送端)进程。

与服务器端相对应,客户端进程包括以下 4 个步骤。

① 创建 DatagramSocket 对象。

除了可以使用上文提到的两个构造方法外,客户端还可以使用如下构造方法。

```
DatagramSocket()
```

创建 DatagramSocket 对象,绑定到本机默认 IP 地址,在本机可用端口中随机指定一个端口。

② 创建 DatagramPacket 对象。

通常,发送方 DatagramPacket 的构造方法为

```
DatagramPacket(byte[] buf, int offset, int length, InetAddress address, int port)
DatagramPacket(byte[] buf, int length, InetAddress address, int port)
```

上述两个构造方法以字节数组 buf 构造 DatagramPacket 对象。其中,数组 buf 存储待发送数据;offset 指定数据报的开始位置;length 表示发送数据的长度;address 与 port 指定接收方的 IP 地址与端口号。

③ 发送 UDP 数据报。

DatagramSocket 类提供如下方法发送数据报。

```
send(DatagramPacket packet)
```

④ 关闭 DatagramSocket 对象。

例 14-2 示范了基于 UDP 的网络 Socket 通信。

例 14-2 服务器(接收)端定义类 Receiver,创建具有固定端口号的 DatagramSocket 对象,侦听来自客户端的数据报。客户端数据报到达之后,接收并显示数据,并向客户端反馈信息。服务器端的代码如下所示。

```java
public class Receiver {
    public static void main(String[] args) throws IOException {
        DatagramSocket s = new DatagramSocket(30000);    //1.创建 DatagramSocket
        byte[] buf = new byte[1024];
        //2.创建 DatagramPacket 对象
```

```
            DatagramPacket pr = new DatagramPacket(buf, buf.length);
            s.receive(pr);              //3.接收数据报
            byte[] bufRev = pr.getData();
            if (bufRev != null && bufRev.length > 0) {
                System.out.println(new String(bufRev, 0, bufRev.length));
                byte[] data = "数据已收到".getBytes();
                DatagramPacket ps = new DatagramPacket(data, data.length,
                                    pr.getAddress(), pr.getPort());          ①
                s.send(ps);             //发送反馈信息
            }
            s.close();                  //4.关闭
        }
    }
```

上述代码行①创建和发送数据报 DatagramPacket 对象,其中接收方的地址与端口号来自数据报对象 pr。

客户(发送)端定义类 Sender,创建具有随机端口号的 DatagramSocket 对象,将用户通过键盘输入的数据封装为数据报并发送到服务器端。客户端最后还需要接收并显示服务器端回送的反馈信息。客户端的代码如下所示。

```
public class Sender {
    public static void main(String[] args) throws IOException {
        DatagramSocket s= new DatagramSocket();     //1.创建 DatagramSocket 对象
        BufferedReader reader = new BufferedReader(
                                    new InputStreamReader(System.in));
        String sendData = reader.readLine();
        byte[] buf = sendData.getBytes();
        //2.封装数据报 DatagramPacket 对象
        DatagramPacket ps = new DatagramPacket(buf,buf.length,
                            InetAddress.getByName("127.0.0.1"), 30000);
        s.send(ps);             //3.发送数据报
        byte[] recBuf = new byte[1024];
        DatagramPacket pr = new DatagramPacket(recBuf,recBuf.length);
        s.receive(pr);          //接收数据报
        System.out.println(new String(pr.getData(),0,pr.getLength()));
        s.close();
    }
}
```

14.4 多线程下载工具程序设计步骤

14.4.1 下载任务实体类

14.4.1

与图 14-1 所示的新建下载任务相对应,构建下载任务实体类,封装下载任务属性。

```java
public class TaskInfo {
    private String URL;                    //下载资源 URL
    private String filename;               //下载后重命名文件名
    private String path;                   //下载后的保存路径
    private int number;                    //定义下载子线程的数目
    public TaskInfo(String URL, String filename, String path, int number) {
        this.URL = URL;
        this.filename = filename;
        this.path = path;
        this.number = number;
    }
}
```

14.4.2 主控界面

1. 主界面

扩展 JFrame，编写类 MultiDownloader 作为程序的主界面。主程序框架中定义文件下载线程 fileDownloadPro、用于显示下载进度情况的 JProgressBar 控件列表及主界面内容面板 contentPane。

```java
public class MultiDownloader extends JFrame{
    private JPanel contentPane;
    private java.util.List<JProgressBar> lstProgress =
                                        new ArrayList<JProgressBar>();
    private FileDownloadPro fileDownloadPro;
    public static void main(String[] args) {
        new MultiDownloader();
    }
    public MultiDownloader() {
        super("下载工具--多线程断点续传");
        JMenuBar menuBar = new JMenuBar();
        JMenu menu = new JMenu("任务");
        menuBar.add(menu);
        JMenuItem newTask = new JMenuItem("新建任务...");
        menu.add(newTask);
        newTask.addActionListener(new ActionListener() {
            @Override
            public void actionPerformed(ActionEvent e) {
                new NewTaskJDialog(MultiDownloader.this,"新建下载任务", true);
            }
        });
        setJMenuBar(menuBar);
        contentPane = new JPanel(new BorderLayout());
```

```java
            getContentPane().add(contentPane);
            setBounds(100, 200, 500, 400);
            setDefaultCloseOperation(JFrame.EXIT_ON_CLOSE);
            setVisible(true);
        }
        //内部类 NewTaskJDialog 的定义
    }
```

2. 新建下载任务对话框 NewTaskJDialog

在主界面可以通过"任务/新建下载任务"子菜单启动新建下载任务的对话框，如图 14-1 所示。新任务对话框 NewTaskJDialog 类定义为主界面 MultiDownloader 的内部类，以便于访问主界面资源，其核心代码如下。

```java
class NewTaskJDialog extends JDialog {
    private JTextField jtfURL = new JTextField(20);      //资源 URL 组件
    private JTextField jtfDir = new JTextField(20);      //保存路径组件
    private JButton btnSelDir = new JButton("...");      //选择路径按钮
    private JTextField jtfName = new JTextField(20);     //文件重命名组件
    private JSpinner spCount = new JSpinner();           //输入子线程数目组件
    private JButton btnOk = new JButton("开始");
    private JButton btnCancle = new JButton("取消");
    public NewTaskJDialog(Frame owner, String title, boolean modal){
        super(owner, title, modal);
        jtfURL.addFocusListener(new FocusAdapter() {
            @Override
            public void focusLost(FocusEvent arg0) {
                String url = jtfURL.getText();
                if (url != null && !url.equals("")) {
                    int index = url.lastIndexOf("/");
                    if (index != -1)
                        //以资源 URL 的文件名作为默认的文件名
                        jtfName.setText(url.substring(index + 1));
                }
            }
        });
        btnSelDir.addActionListener(new ActionListener() {
            @Override
            public void actionPerformed(ActionEvent e) {
                File current = new File(NewTaskJDialog.class
                                    .getResource("/").getPath());
                JFileChooser chooser = new JFileChooser(current);
                chooser.setFileSelectionMode(JFileChooser.DIRECTORIES_ONLY);
                int approval =
                        chooser.showOpenDialog(NewTaskJDialog.this);
```

```java
            if (approval = = JFileChooser.APPROVE_OPTION) {
                File target = chooser.getSelectedFile();
                jtfDir.setText(target.getAbsolutePath());      //选择保存路径
            }
        }
    });
    btnCancle.addActionListener(new ActionListener() {
        @Override
        public void actionPerformed(ActionEvent e) {
            NewTaskJDialog.this.dispose();
        }
    });
    btnOk.addActionListener(new ActionListener() {
        @Override
        public void actionPerformed(ActionEvent e) {
            initDownloadProcPal();              //初始化下载进度面板
            NewTaskJDialog.this.dispose();
            download();                         //开始下载
        }
    });
}//NewTaskJDialog 类构造方法结束
//定义 initDownloadProcPal 方法
//定义 download 方法
}//内部类 NewTaskJDialog 定义结束
```

3. 初始化下载进度面板 initDownloadProcPal( )

NewTaskJDialog 类中定义 initDownloadProcPal( )方法，根据新建的下载任务初始化主界面的内容面板 contentPane 对象。为了更好地模拟断点续传功能，添加"暂停""继续"这两个命令按钮，界面效果如图 14-2 所示。

```java
public void initDownloadProcPal() {
    JLabel lblName = new JLabel("正在下载的文件: " + jtfName.getText());
    JButton btnPause = new JButton("暂停");
    JButton btnResume = new JButton("继续");
    btnResume.setEnabled(false);
    btnPause.addActionListener(new ActionListener() {
        @Override
        public void actionPerformed(ActionEvent arg0) {
            btnResume.setEnabled(true);
            btnPause.setEnabled(false);
            fileDownloadPro.setStop(true);        //结束下载线程
        }
    });
    btnResume.addActionListener(new ActionListener() {
```

```java
            @Override
            public void actionPerformed(ActionEvent e) {
                btnPause.setEnabled(true);
                btnResume.setEnabled(false);
                download();                          //开始下载
            }
        });
        JPanel top = new JPanel();
        top.add(lblName);
        top.add(btnPause);
        top.add(btnResume);
        int proNumber = (int) spCount.getValue();
        JLabel lblSplit = new JLabel("分为" + proNumber + "线程：");
        top.add(lblSplit);
        contentPane.removeAll();
        contentPane.add(top, BorderLayout.NORTH);
        JPanel process = new JPanel();
        process.setLayout(new BoxLayout(process, BoxLayout.Y_AXIS));
        lstProgress.clear();
        for (int i = 0; i < proNumber; i++) {
            JProgressBar bar = new JProgressBar(JProgressBar.HORIZONTAL, 0, 100);
            bar.setSize(300, 10);
            bar.setStringPainted(true);
            Box line = Box.createHorizontalBox();
            line.add(new JLabel("线程" + (i + 1)));
            line.add(bar);
            process.add(line);
            process.add(Box.createVerticalStrut(10));
            lstProgress.add(bar);
        }
        contentPane.add(process);
        contentPane.validate();
}//initDownloadProcPal 方法定义结束
```

4. 下载文件 download()方法

NewTaskJDialog 类定义 download()方法，用来构建新的下载任务，启动文件下载线程。

```java
public void download() {
    try {
        TaskInfo target = new TaskInfo(jtfURL.getText(),
                        jtfName.getText(), jtfDir.getText(),
                        (int) spCount.getValue());
        fileDownloadPro = new FileDownloadPro(target, lstProgress);
```

```
            fileDownloadPro.start();
        } catch (Exception e) {
            e.printStackTrace();
        }
    }//download()方法定义结束
```

14.4.3 文件下载线程

1. 类 FileDownloadPro 的定义

文件下载主线程 FileDownloadPro 是 Swing GUI 的工作线程，负责启动若干文件下载子线程，并提供保存断点与断点续传功能。

```
public class FileDownloadPro extends Thread{
    private TaskInfo task;                        //下载任务
    private List<JProgressBar> lstProgress;       //下载进度展示组件
    private File tempFile;                        //临时文件,用于保存下载断点信息
    private FileDownSubPro[] subPros;             //下载子线程数组
    private long[] start;                         //子线程下载起始位置数组
    private long[] end;                           //子线程下载截止位置数组
    boolean isStop = false;                       //标记线程是否终止
    //构造方法
    //定义线程执行体
    //定义方法,读取断点信息
    //定义方法,保存断点信息
}
```

2. 构造方法

FileDownloadPro 类构造方法的主要功能是根据下载任务生成下载子线程数组 subPros、子线程位置数组 start 与 end,其核心代码如下。

```
public FileDownloadPro(TaskInfo task, List<JProgressBar> lstProgress)
                                         throws IOException {
    this.task = task;
    this.lstProgress = lstProgress;
    tempFile = new File(task.getPath() + File.separator
                        + task.getFilename() + ".tmp");
    if (tempFile.exists()) {
        readBreakPoint();         //读取断点信息
        subPros = new FileDownSubPro[start.length];
    } else {                       //新文件下载
        URL url = new URL(task.getURL());
        HttpURLConnection httpCon = (HttpURLConnection) url.openConnection();
        long fileLength = httpCon.getContentLength();
```

```java
        if (-1 == fileLength) {
            JOptionPane.showMessageDialog(null, "文件的长度未知", ""
                                            , JOptionPane.ERROR_MESSAGE);
            return;
        }
        subPros = new FileDownSubPro[task.getNumber()];
        start = new long[subPros.length];
        end = new long[subPros.length];
        long blockLength = fileLength/task.getNumber();
        for (int i = 0; i < subPros.length; i++)
            start[i] = i * blockLength;
        for (int i = 0; i < subPros.length -1; i++)
            end[i] = start[i + 1];
        end[subPros.length - 1] = fileLength;
    }
}//FileDownloadPro 构造方法定义结束
```

3. 线程执行体

文件下载线程首先启动若干个下载子线程,实现网络资源的下载,然后侦听下载线程的中断标记,如果界面中断下载线程的执行,则保存断点信息,以备断点续传;反之,如果界面没有中断文件下载线程的执行,则在各个子线程下载结束后删除临时文件。

```java
public void run() {
    try {
        for (int i = 0; i < subPros.length; i++ ) {     //创建并启动线程
            subPros[i] = new FileDownSubPro(task, start[i], end[i]
                                        , lstProgress.get(i));
            subPros[i].start();
        }
        boolean isFinished = false;
        while (!isStop) {
            isFinished = true;                          //判断下载是否结束
            for (int i = 0; i < subPros.length; i++) {
                if (subPros[i].isAlive()){
                    isFinished = false;
                    break;
                }
            }
            if (isFinished) {
                JOptionPane.showMessageDialog(null, "下载结束");
                if (tempFile.exists()) {
                    tempFile.delete();                  //删除临时文件
                }
                break;
```

```
            }
        }
        if (!isFinished) {                              //保存断点
            for (int i = 0; i < subPros.length; i++)
                subPros[i].setStop(true);
            writeBreakPoint();
        }
    } catch (Exception e) {
        e.printStackTrace();
    }
}//线程执行体定义结束
```

4. 读取与保存断点信息

FileDownloadPro 类定义 writeBreakPoint()方法,将各下载子线程的下载位置作为线程断点信息保存至临时文件。

同理,定义 readBreakPoint()方法,从临时文件中读取断点信息,以备续传。

```
private void writeBreakPoint() {       //保存断点信息
    try {
        DataOutputStream output = new DataOutputStream(
                         new FileOutputStream(tempFile));
        List<Long> start = new ArrayList<Long>();
        List<Long> end = new ArrayList<Long>();
        for (int i = 0; i < subPros.length; i++) {
            if (subPros[i].getCurLocation() < subPros[i].getEnd()) {
                start.add(subPros[i].getCurLocation());
                end.add(subPros[i].getEnd());
            }
        }
        output.writeInt(start.size());
        for (int i = 0; i < start.size(); i++) {
            output.writeLong(start.get(i));
            output.writeLong(end.get(i));
        }
        output.close();
    } catch (Exception e) {
        e.printStackTrace();
    }
}//writeBreakPoint 方法定义结束
private void readBreakPoint() {        //读取断点信息
    try {
        DataInputStream input = new DataInputStream(
                         new FileInputStream(tempFile));
        int count = input.readInt();
```

```
            start = new long[count];
            end = new long[count];
            for (int i = 0; i < count; i++) {
                start[i] = input.readLong();
                end[i] = input.readLong();
            }
            input.close();
        } catch (Exception e) {
            e.printStackTrace();
        }
    }//readBreakPoint 方法定义结束
```

14.4.4 文件下载子线程

14.4.4

1. 类 FileDownSubPro 的定义

文件下载子线程负责网络资源的分段下载。

```
public class FileDownSubPro extends Thread {
    private TaskInfo task;                      //下载任务
    private JProgressBar bar;                   //与当前子线程对应的进度条控件
    private long curLocation;                   //子线程读取的当前位置
    private long end;                           //子线程读取的结束位置
    private boolean isStop = false;             //标记子线程是否终止
    public FileDownSubPro(TaskInfo task,long curLocation
                                        ,long end,JProgressBar bar) {
        this.task = task;
        this.curLocation = curLocation;
        this.end = end;
        this.bar = bar;
    }
    //线程体 run()方法的定义
}
```

2. 多线程并发与 Swing GUI 组件的动态刷新

　　文件下载子线程需要根据下载进度更新主界面的 JProgressBar 控件。由于 Java Swing 是单线程、线程不安全的图形库，所以通过多线程对用户界面中的 Swing 组件直接进行刷新操作会造成用户界面的刷新失败。

　　虽然 Swing 的 API 不是线程安全的，但是 Swing 框架采用事件分发线程(Event Dispatch Thread, EDT)的方式保证线程安全。Swing 框架中定义了一个先进先出的事件队列(Event Queue)，GUI 界面上发出的请求事件(如窗口移动、点击按钮及组件的重绘等)都被添加到该事件队列。添加到该队列的任务将按顺序逐一在同一线程中被执行，该线程被称为事件分发线程。所以要想构建线程安全的 Swing GUI 程序，必须保证将刷新 GUI 组件的操作以任务的形式添加到事件队列，最终在事件分发线程中被调用。Java.swing.SwingUtilities 类提供如下实现添加任务到事件队列的静态方法。

```
void invokeLater(Runnable doRun)
void invokeAndWait(Runnable doRun)
```

invokeLater()方法是异步的，它会立即返回，具体何时执行任务 doRun 并不确定；invokeAndWait()方法是同步的，它在被调用结束时会立即阻塞当前线程，直到事件分发线程处理完 doRun 任务。

在文件下载子线程的执行体中实现了文件下载与下载进度更新操作，核心代码如下。

```java
public void run() {
    RandomAccessFile randomFile = null;
    try {
        randomFile = new RandomAccessFile(task.getPath()
                + File.separator + task.getFilename(), "rw");
        randomFile.seek(getCurLocation());
        URL url = new URL(task.getURL());
        HttpURLConnection httpConn = (HttpURLConnection) url.openConnection();
        httpConn.setRequestMethod("GET");
        //设置请求数据的区间
        httpConn.setRequestProperty("Range", "bytes= "
                        + getCurLocation() + "-" + getEnd());
        InputStream input = httpConn.getInputStream();
        byte[] b = new byte[1024];
        int readNum;
        long start = getCurLocation();
        while ((readNum = input.read(b, 0, 1024)) > 0 &&
            this.getCurLocation() < this.getEnd() && !isStop) {
            randomFile.write(b, 0, readNum);
            setCurLocation(getCurLocation() + readNum);
            double process = 1.0 * (getCurLocation() - start)
                                            / (getEnd() - start);
            SwingUtilities.invokeLater(new Runnable() {
                @Override
                public void run() {
                    bar.setValue((int) (process * 100));
                    bar.repaint();
                }
            });
        }
    } catch (Exception e) {
        e.printStackTrace();
    } finally {
        try {
```

```
                if (randomFile != null)
                    randomFile.close();
            } catch (IOException e) {
                e.printStackTrace();
            }
        }
    }//run()方法定义结束
```

14.5 练一练

利用基于 TCP 的 Socket 编写一个聊天室程序,要求具备以下功能。

1. 服务器端。

(1) 显示本机的 IP 地址,管理员输入侦听端口,启动服务器。

(2) 可以接收来自于多个客户端的连接,并显示当前在线的客户名称。

(3) 分别以一对一和一对多的方式实现客户端消息的转发。

(4) 显示转发的消息及发送和接收消息的客户名称。

2. 客户端。

(1) 输入服务器端的 IP 地址,端口号及客户名称,连接服务器。

(2) 根据服务器端传送的客户列表选择一对一或者群发消息。

(3) 显示传送到本客户端的消息及发送和接收消息的客户名称。

第 15 章 房屋租赁系统的设计与实现

本章将根据项目的需求规格说明书及体系结构设计文档,运用基本的面向对象技术完成一个房屋租赁系统的实现、测试及发布。通过本章综合项目的训练,读者能够掌握利用 Java 语言实现面向对象系统开发的方法、技术及过程。

本章的主要内容包括:
- 软件需求规格说明;
- 软件体系结构设计规格说明;
- 系统实现;
- 系统测试;
- 程序发布。

15.1 软件需求规格说明

15.1.1 总体描述

房屋租赁系统的主要角色为房屋租赁公司前台用户,包括如下功能。
(1) 登录。
(2) 房屋管理:房屋的增、删、改、查及报表输出操作。
(3) 客户管理:客户的增、改、查及报表输出操作。
(4) 求租意向:承租人求租意向的增、删、改、查及报表输出操作。
(5) 合同管理:签订合同、提前解约、查询及报表输出操作。
(6) 业务员管理:业务人员入职、离职、修改、查询及报表输出操作。

房屋租赁系统的功能结构如图 15-1 所示。

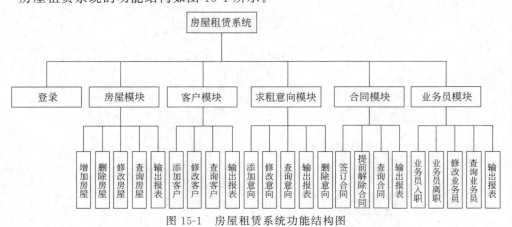

图 15-1 房屋租赁系统功能结构图

15.1.2 具体需求

1. 原型系统界面及数据规范

限于篇幅,下面仅以房屋模块中的"新增房屋"为例,描述原型系统界面及相关数据规范。

窗口标题:房屋管理。

页面布局如图 15-2 所示。

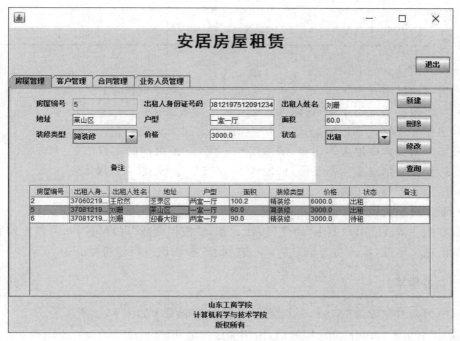

图 15-2 房屋管理原型界面

数据格式及有效范围如表 15-1 所示。

表 15-1 房屋数据格式及有效范围

名称	类型	有效范围	度量单位	缺省值	正则表达式	可否为空	I/O
房屋编号	int	1~1000000	NA	无		否	I/O
出租人身份证号码	String	18 位字母、数字	NA	无	[a-zA-Z0-9]{18}	否	I/O
出租人姓名	String	1~10 位字母或汉字	NA	无	[\u4e00-\u9fa5a-zA-Z]{1,10}	否	I/O
地址	String	1~255 位字母、数字或汉字	NA	无	[\u4e00-\u9fa5a-zA-Z0-9]{1,255}	否	I/O

续表

名称	类型	有效范围	度量单位	缺省值	正则表达式	可否为空	I/O
房型	String	1~10位字母或汉字	NA	无	[\u4e00-\u9fa5a-zA-Z]{1,10}	可	I/O
面积	float	0~1000.0	平方米	无		否	I/O
装修类型	String	精装修、简装修、毛坯房	NA	无		可	I/O
价格	float	0~100000.0	元	无		否	I/O
状态	String	待租、出租	NA	无		否	I/O
备注	String	1~255位字母、数字或汉字	NA	无	[\u4e00-\u9fa5a-zA-Z0-9]{1,255}	可	I/O

窗口反馈信息如下：
- 房屋地址格式不正确，请输入1~255位字母、数字或汉字；
- 房型数据格式不正确，请输入1~10位字母或汉字；
- 房屋面积格式不正确，请输入0~1000.0之间的实数；
- 房屋租赁价格输入不正确，请输入0~100000.0之间的实数；
- 房屋备注信息输入不正确，请输入1~255位字母、数字或汉字；
- 出租人信息不正确；
- 新建房屋失败。

命令执行方式：用户单击"新建"按钮，"新建"按钮变成"保存"按钮，并清空屏幕上方的房屋信息文本框，输入房屋信息后，单击"保存"按钮执行新建房屋命令。

2. 功能需求

采用面向对象原理的use case描述，系统的功能需求如图15-3所示。

以"新增房屋（AddHouse）"为例，用例描述如表15-2所示。

表15-2 AddHouse 用例描述

用例标识符	AddHouse		
用例名称	添加房屋信息		
用例创建者	于立萍	用例的最后修改者	于立萍
用例创建时间	2020.9.8	用例的最后修改时间	2020.9.8
操作者	用户		
描述	用户添加房屋详细信息		
前置条件	系统从房屋信息表（HouseInformationTable）中查找到房屋信息，形成房屋信息实例列表（HouseInformationList）		
后置条件	系统将用户输入的房屋信息保存到房屋信息表（HouseInformationTable），显示房屋管理界面（HouseManageGUI）		

第15章 房屋租赁系统的设计与实现

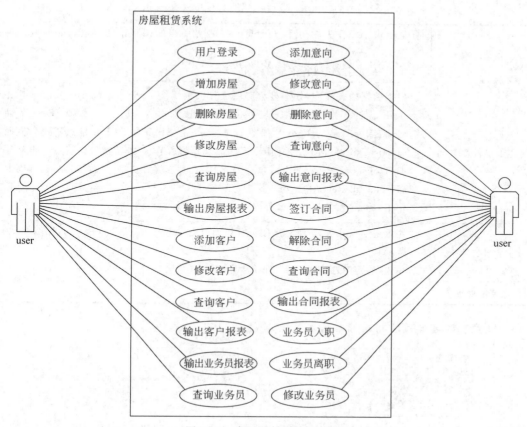

图15-3 房屋租赁系统用例图

续表

	用户	系统
主事件流		① 根据房屋信息实例列表(HouseInformationList)显示房屋管理界面(HouseManageGUI)
	② 用户单击"新建"按钮	③ 清空房屋信息输入框,"新建"按钮变更为"保存"按钮
	④ 用户输入房屋信息,单击"保存"按钮	⑤ 判断输入的房主信息是否正确,若不正确,则执行子事件流 a;判断其他的房屋信息格式是否符合规范,若不符合规范,则执行子事件流 b
		⑥ 将用户输入的房屋信息保存到房屋信息表(HouseInformationTable)。若保存失败,则执行子事件流 c
		⑦ 从房屋信息表(HouseInformationTable)中找到房屋信息,形成房屋信息实例列表(HouseInformationList),刷新房屋管理界面(HouseManageGUI)

261

续表

子事件流 a	① 显示房屋操作失败-出租人信息不正确界面（HouseFail-HouseOwnerError）如图15-4所示 ② 单击"确定"按钮执行主事件流 4
子事件流 b	① 根据数据格式标准，分别显示房屋操作失败-地址信息不正确界面（HouseFail-AddressError），如图 15-5 所示；房屋操作失败-房型信息不正确界面（HouseFail-LayoutError），如图 15-6 所示；房屋操作失败-面积信息不正确界面（HouseFail-AreaError），如图 15-7 所示；房屋操作失败-租赁价格信息不正确界面（HouseFail-PriceError），如图 15-8 所示；房屋操作失败-备注信息不正确界面（HouseFail-RemarkError），如图 15-9 所示。 ② 单击"确定"按钮执行主事件流 4
子事件流 c	① 显示房屋操作失败-新建房屋失败界面（HouseFail-AddHouseError），如图 15-10 所示 ② 单击"确定"按钮执行主事件流 4
异常处理	

注意：数据格式的判断标准请参见表15-1的说明。

图 15-4　房屋操作失败-出租人信息不正确

图 15-5　房屋操作失败-地址信息不正确

图 15-6　房屋操作失败-房型信息不正确

图 15-7　房屋操作失败-面积信息不正确

图 15-8　房屋操作失败-租赁价格信息不正确

图 15-9　房屋操作失败-备注信息不正确

图 15-10　房屋操作失败-新建房屋失败

3. 性能需求

用户操作软件完成后，系统会给出用户反馈。表的增加、删除、修改、查找等操作在 2 秒内完成，在 1000 个用户并发操作的环境下，要求软件的延迟时间不超过 4 秒。

4. 数据库需求

本项目支持数据库的增加、删除、修改和查找等基本操作，可以对字符串、数字等基本数据类型的数据进行存储。数据实体及其关系如图 15-11 所示。

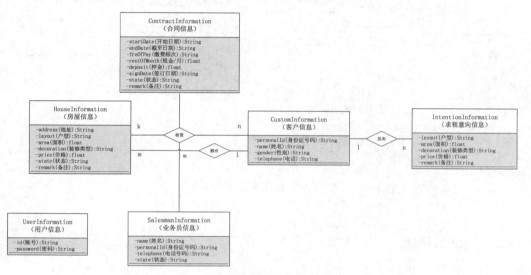

图 15-11 房屋租赁系统数据实体及关系

15.2 体系结构设计

本节将利用 UML 图描述系统体系结构的不同视图。

15.2.1 Use Case 实现

以"新增房屋"为例，其用例实现如图 15-12 所示。

① CustomService::searchCustomByPersonalId(String personalId)
　　　　　　　　　　　　　　　　　　　　　　　　　　:CustomInformation
② CustomDao::searchCustomByPersonalId(String personalId)
　　　　　　　　　　　　　　　　　　　　　　　　　　:CustomInformation
③ HouseService::addHouse(HouseInformation houseInfo):boolean
④ HouseService::searchAllHouses():List<HouseInformation>
⑤ HouseDao::addHouse(HouseInformation houseInfo):int
⑥ HouseDao::searchAllHouses():List<HouseInformation>

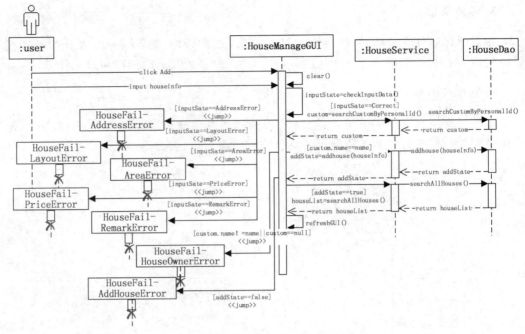

图 15-12 添加房屋设计序列图

15.2.2 逻辑视图

利用层次化的包或子系统描述系统模型的分解，系统包图如图 15-13 所示。

房屋租赁系统采用 C/S 模式，共分为 6 层，分别为 UserInterface（用户交互层），BusinessService（业务逻辑接口层），BusinessServiceImpl（业务逻辑层），DatabaseAccess（数据访问接口层），DatabaseAccessImpl（数据访问层），Bean（数据实体层）。

15.2.3 部署视图

利用部署图描述系统的网络配置（物理节点）。系统部署视图如图 15-14 所示。

15.2.4 实现视图

针对逻辑视图（如图 15-13）中定义的类进行详细说明。限于篇幅，下面以"新增房屋"功能为例进行实现说明。

1. 业务服务层

以 HouseServiceImpl 为例，类图如图 15-15 所示。

第15章 房屋租赁系统的设计与实现

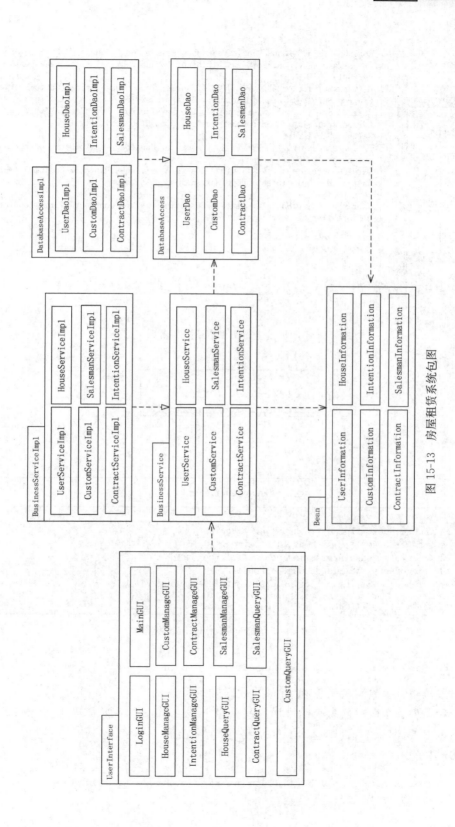

图 15-13 房屋租赁系统包图

图 15-14　房屋租赁系统部署图

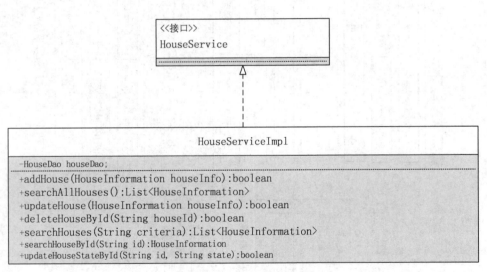

图 15-15　HouseServiceImpl 类图

　　HouseServiceImpl::addHouse(HouseInformation houseInfo):boolean

　　前置条件：参数满足表 15-1 所示的数据格式。
　　后置条件：在房屋信息表中添加房屋信息 houseInfo，若添加成功，则更新 houseInfo 的房屋 Id 并返回 true，否则返回 false。
　　伪码如下：

　　HouseServiceImpl::addHouse(HouseInformation houseInfo):boolean

```
{
    id←houseDao.addHouse(houseInfo);
    if(id!=-1){
        houseInfo.setId(id);
        return true;
    } else
        return false;
}
```

2. 数据访问层

以 HouseDaoImpl 为例，类图如图 15-16 所示。

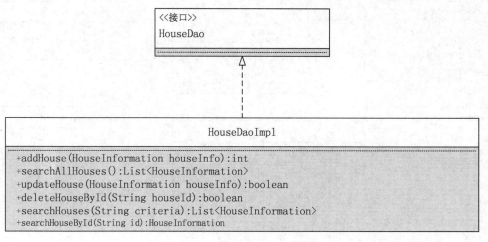

图 15-16　HouseDaoImpl 类图

HouseDaoImpl::addHouse(HouseInformation houseInfo):int

前置条件：参数满足表 15-1 所示的数据格式。

后置条件：在房屋信息表中添加房屋信息 houseInfo，若添加成功，则返回生成的房屋 Id，否则返回−1。

伪码如下：

```
HouseDaoImpl::addHouse(HouseInformation houseInfo):int
{
    id←-1;
    insertLineNumber ← insert into HouseInformationTable (ownerId, address, layout, area, decoration, price, state, remark) values (houseInfo.ownerId, houseInfo.address, houseInfo.layout, houseInfo.area, houseInfo.decoration, houseInfo.price, houseInfo.state, houseInfo.remark);
    if(insertLineNumber == ONE_LINE){
        id←getGeneratedKeys();
    }
```

```
    return id;
}
```

15.2.5 数据视图

房屋租赁系统的逻辑数据模型如图 15-17 所示。

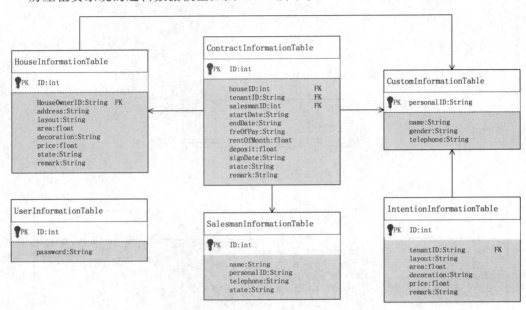

图 15-17 房屋租赁系统逻辑数据模型

1. UserInformationTable

UserInformationTable 用来描述用户信息,逻辑设计如表 15-3 所示。

表 15-3 UserInformationTable

名称	类型	长度限制	缺省值	可否缺省	可否为空	PK/FK
ID	int		无	否	否	PK
password	String	20	无	否	否	

2. HouseInformationTable

HouseInformationTable 用来描述房屋信息,逻辑设计如表 15-4 所示。

3. CustomInformationTable

CustomInformationTable 用来描述客户信息,逻辑设计如表 15-5 所示。

表 15-4　HouseInformationTable

名称	类型	长度限制	缺省值	可否缺省	可否为空	PK/FK
ID	int		无	否	否	PK
HouseOwnerID	String	18	无	否	否	FK
Address	String	255	无	否	否	
Layout	String	10	无	可	可	
Area	float		无	否	否	
Decoration	String	3	无	可	可	
Price	float		无	否	否	
State	String	2	无	否	否	
Remark	String	255	无	可	可	

表 15-5　CustomInformationTable

名称	类型	长度限制	缺省值	可否缺省	可否为空	PK/FK
PersonalID	String	18	无	否	否	PK
Gender	String	1	无	可	可	
name	String	10	无	否	否	
Telephone	String	13	无	否	否	

4. ContractInformationTable

ContractInformationTable 用来描述房屋合同信息，逻辑设计如表 15-6 所示。

表 15-6　ContractInformationTable

名称	类型	长度限制	缺省值	可否缺省	可否为空	PK/FK
ID	int		无	否	否	PK
HouseID	int		无	否	否	FK
TenantID	String	18	无	否	否	FK
SalesmanID	int		无	否	否	FK
StartDate	String	10	无	否	否	
EndDate	String	10	无	否	否	
FreOfPay	String	2	无	否	否	
RentOfMonth	float		无	否	否	
Deposit	float		无	否	否	
SignDate	String	10	无	否	否	
State	String	4	无	否	否	
Remark	String	255	无	可	可	

5. SalesmanInformationTable

SalesmanInformationTable 用来描述销售人员信息，逻辑设计如表 15-7 所示。

表 15-7　SalesmanInformationTable

名称	类型	长度限制	缺省值	可否缺省	可否为空	PK/FK
ID	int		无	否	否	PK
PersonalID	String	18	无	否	否	
Name	String	10	无	否	否	
Telephone	String	13	无	否	否	
State	String	2	无	否	否	

6. IntentionInformationTable

IntentionInformationTable 用来描述客户的求租意向，逻辑设计如表 15-8 所示。

表 15-8　IntentionInformationTable

名称	类型	长度限制	缺省值	可否缺省	可否为空	PK/FK
ID	int		无	否	否	PK
TenantID	String	18	无	否	否	FK
Layout	String	10	无	可	可	
Area	float		无	可	可	
Decoration	String	3	无	可	可	
Price	float		无	可	可	
Remark	String	255	无	可	可	

15.3　编码实现

15.3.1　项目的文件结构

根据体系结构设计规格说明，项目文件结构如图 15-18 所示。

15.3.2

15.3.2　数据实体包 bean

项目涉及以下 8 个数据实体类（限于篇幅，省略类的部分方法定义，完整内容请参见本书的附代资源）。

第15章 房屋租赁系统的设计与实现

图 15-18　房屋租赁系统项目结构

1. 用户 UserInformation

```
public class UserInformation {
    private int id;
    private String password;
}
```

2. 合同 ContractInformation

```
public class ContractInformation {
    private int id;
    private int houseId;
    private String houseAddress;
    private String ownerId;
    private String ownerName;
    private String tenantId;
    private String tenantName;
    private int salesmanId;
    private String salesName;
    private String startDate;
    private String endDate;
    private String freOfPay;
    private float rentOfMonth;
    private float deposit;
    private String signDate;
    private String state;
    private String remark;
}
```

3. 合同 ContractDB

```
public class ContractDB {
    private int id;
```

271

```java
        private int houseId;
        private String tenantId;
        private int salesmanId;
        private String startDate;
        private String endDate;
        private String freOfPay;
        private float rentOfMonth;
        private float deposit;
        private String signDate;
        private String state;
        private String remark;
        public ContractDB() {
        }
}
```

4. 客户 CustomInformation

```java
public class CustomInformation {
    private String personalId;
    private String name;
    private String gender;
    private String telephone;
}
```

5. 房屋 HouseInformation

```java
public class HouseInformation {
    private int id;
    private String ownerId;
    private String ownerName;
    private String address;
    private String layout;
    private float area;
    private String decoration;
    private float price;
    private String state;
    private String remark;
    public HouseInformation() {
    }
}
```

6. 房屋 HouseDB

```java
public class HouseDB {
    private int id;
    private String  ownerId;
```

```java
    private String address;
    private String layout;
    private float area;
    private String decoration;
    private float price;
    private String state;
    private String remark;
}
```

7. 求租意向 IntentionInformation

```java
public class IntentionInformation {
    private int id;
    private String tenantId;
    private String layout;
    private float area;
    private String decorate;
    private float price;
    private String remark;
    public IntentionInformation() {
    }
}
```

8. 业务人员 SalesmanInformation

```java
public class SalesmanInformation {
    private int id;
    private String name;
    private String personalId;
    private String telephone;
    private String state;
}
```

15.3.3 数据访问包 dao

1. 接口 UserDao 及其实现类 UserDaoImpl

```java
public interface UserDao {
    public UserInformation login(int id, String pwd);
}
public class UserDaoImpl implements UserDao {
    @Override
    public UserInformation login(int id, String pwd) {
        String sql="select * from user where id=?";
        List<Object> params=new ArrayList<Object>();
```

15.3.3

```
        params.add(id);
        List<UserInformation> list = DBTool.executeQuery(UserInformation.class,
        sql,params);
        if(list!=null && list.size()>0)
            return list.get(0);
        else
            return null;
    }
}
```

2. 接口 HouseDao 及其实现类 HouseDaoImpl

```
public interface HouseDao {
    List<HouseInformation> searchAllHouses();
    List<HouseInformation> searchHouses(QueryCriteria criteria, List<Object>
                                        opt);
    HouseInformation searchHouseById(String id);
    boolean deleteHouseById(String id);
    int addHouse(HouseDB h) throws SQLException;
    boolean updateHouse(HouseDB h) throws SQLException;
}
public class HouseDaoImpl implements HouseDao {
    @Override
    public int addHouse(HouseDB h) throws SQLException {
        int id=-1;
        Connection conn=DBTool.getConnection();
        String sql="insert into house(ownerId,address,layout,area,
            decoration,price,state,remark) values (?,?,?,?,?,?,'待出租',?)";
        PreparedStatement pst=conn.prepareStatement(sql,
                                        Statement.RETURN_GENERATED_KEYS);
        pst.setString(1, h.getOwnerId());
        pst.setString(2, h.getAddress());
        pst.setString(3, h.getLayout());
        pst.setFloat(4, h.getArea());
        pst.setString(5, h.getDecoration());
        pst.setFloat(6, h.getPrice());
        pst.setString(7, h.getRemark());
        int result=pst.executeUpdate();
        if(result==1) {
            ResultSet keys=pst.getGeneratedKeys();
            keys.next();
            id=keys.getInt(1);
        }
        pst.close();
        return id;
```

```java
    }
    @Override
    public boolean updateHouse(HouseDB h) throws SQLException {
        boolean result=false;
        Connection conn=DBTool.getConnection();
        String sql="update house set ownerId=?,address=?,layout=?,area=?
                    ,decoration=?,price=?,state=?,remark=? where id=?";
        PreparedStatement pst=conn.prepareStatement(sql);
        pst.setString(1, h.getOwnerId());
        pst.setString(2, h.getAddress());
        pst.setString(3, h.getLayout());
        pst.setFloat(4, h.getArea());
        pst.setString(5, h.getDecoration());
        pst.setFloat(6, h.getPrice());
        pst.setString(7, h.getState());
        pst.setString(8, h.getRemark());
        pst.setInt(9, h.getId());
        int count=pst.executeUpdate();
        if(count==1) {
            result=true;
        }
        pst.close();
        return result;
    }
    @Override
    public List<HouseInformation> searchAllHouses() {
        String sql="select house.*,custom.name ownerName from house left join
        custom on house.ownerId=custom.personalId where house.state!='已删
        除'";
        return DBTool.executeQuery(HouseInformation.class, sql, null);
    }
    @Override
    public List<HouseInformation> searchHouses(QueryCriteria criteria, List
                                        <Object> opt) {
        String sql="select house.*,custom.name ownerName from house
                left join custom on house.ownerId=custom.personalId
                    where house.state!='已删除' and "+criteria.getCriteria();
        return DBTool.executeQuery(HouseInformation.class, sql, opt);
    }
    @Override
    public HouseInformation searchHouseById(String id) {
        String sql="select house.*,custom.name ownerName from house
                left join custom on house.ownerId=custom.personalId
                    where house.state!='已删除' and house.id="+id;
```

```java
        List<HouseInformation> list=DBTool.executeQuery (
                                    HouseInformation.class, sql, null);
        if(list !=null && list.size()> 0)
            return list.get(0);
        else
            return null;
    }
    @Override
    public boolean deleteHouseById(String id) {
        boolean result=false;
        String sql="select * from house where id=?";
        List<Object> params=new ArrayList<Object> ();
        params.add(id);
        List<HouseDB> list=DBTool.executeQuery(HouseDB.class, sql, params);
        if(list !=null) {
            HouseDB h=list.get(0);
            h.setState("已删除");
            try {
                result=updateHouse(h);
            } catch (SQLException e) {
                e.printStackTrace();
            }
        }
        return result;
    }
}
```

3. 接口 CustomDao 及其实现类 CustomDaoImpl

```java
public interface CustomDao {
    boolean addCustom(CustomInformation o) throws SQLException;
    boolean updateCustom(CustomInformation o) throws SQLException;
    public List<CustomInformation>searchAllCustoms();
    public List<CustomInformation>searchCustoms(QueryCriteria criteria, List
    <Object>opt);
    public CustomInformation searchCustomByPersonalID(String pID);
}
public class CustomDaoImpl implements CustomDao {
    @Override
    public boolean addCustom(CustomInformation o) throws SQLException {
        Connection conn=DBTool.getConnection();
        String sql="insert into custom(personalId,name,gender,telephone)
        values (?,?,?,?)";
        PreparedStatement pst=conn.prepareStatement(sql);
        pst.setString(1, o.getPersonalId());
```

```java
            pst.setString(2, o.getName());
            pst.setString(3, o.getGender());
            pst.setString(4, o.getTelephone());
            int result=pst.executeUpdate();
            if(result==1) {
                return true;
            }
            pst.close();
            return false;
        }
        @Override
        public boolean updateCustom(CustomInformation o) throws SQLException {
            boolean result=false;
            Connection conn=DBTool.getConnection();
            String sql="update custom set name=?,gender=?,telephone=? where personalId=?";
            PreparedStatement pst=conn.prepareStatement(sql);
            pst.setString(4, o.getPersonalId());
            pst.setString(1, o.getName());
            pst.setString(2, o.getGender());
            pst.setString(3, o.getTelephone());
            int count=pst.executeUpdate();
            if(count==1) {
                result=true;
            }
            pst.close();
            return result;
        }
        @Override
        public List<CustomInformation>searchAllCustoms() {
            String sql="select personalId,name,gender,telephone from custom ";
            return DBTool.executeQuery(CustomInformation.class, sql, null);
        }
        @Override
        public List<CustomInformation>searchCustoms(QueryCriteria criteria, List
                                                                <Object>opt) {
            String sql="select * from Custom where "+criteria.getCriteria();
            return DBTool.executeQuery(CustomInformation.class, sql, opt);
        }
        @Override
        public CustomInformation searchCustomByPersonalID(String pID) {
            String sql="select name,gender,telephone from custom where personalId=?";
            List<Object>params=new ArrayList<Object>();
```

```java
        params.add(pID);
        List<CustomInformation>list=DBTool.executeQuery (
                            CustomInformation.class, sql, params);
        if(list!=null && list.size()>0)
            return list.get(0);
        else
            return null;
    }
}
```

4. 接口 ContractDao 及其实现类 ContractDaoImpl

```java
public interface ContractDao {
    List<ContractInformation>searchAllContracts();
    List<ContractInformation>searchContracts(QueryCriteria criteria, List
                                        <Object>opt);
    int addContract(ContractDB c) throws SQLException;
    boolean updateContract(ContractDB c) throws SQLException;
}
public class ContractDaoImpl implements ContractDao {
    static String sql="select contract.id,contract.houseId
        ,house.address houseAddress,house.ownerid ownerId
        , contract.tenantId, custom.name tenantName,contract.salesmanId
        ,salesman.name salesName, contract.startDate, ,contract.endDate
        ,contract.freOfPay,contract.rentOfMonth,contract.deposit
        ,contract.signDate, contract.state,contract.remark
            from ((((contract left join house on contract.houseId=house.id)
                left join custom on custom.personalId=contract.tenantId)
                left join custom c on c.personalId=house.ownerID)
                left join salesman on salesman.id=contract.salesmanId)";
    @Override
    public int addContract(ContractDB c) throws SQLException {
        int id=-1;
        Connection conn=DBTool.getConnection();
        String sql="insert into contract(houseId,tenantId,salesmanId, startDate
                ,endDate,freOfPay, rentOfMonth,deposit,signDate,state,remark)
                values (?,?,?,?,?,?,?,?,?,?,?)";
        SimpleDateFormat sdf=new SimpleDateFormat("yyyy-MM-dd");
        String signDate=sdf.format(new Date());
        PreparedStatement pst=conn.prepareStatement(sql
                                ,Statement.RETURN_GENERATED_KEYS);
        pst.setInt(1, c.getHouseId());
        pst.setString(2, c.getTenantId());
        pst.setInt(3, c.getSalesmanId());
        pst.setString(4, c.getStartDate());
```

```java
        pst.setString(5, c.getEndDate());
        pst.setString(6, c.getFreOfPay());
        pst.setFloat(7, c.getRentOfMonth());
        pst.setFloat(8, c.getDeposit());
        pst.setString(9, signDate);
        pst.setString(10, c.getState());
        pst.setString(11, c.getRemark());
        int result=pst.executeUpdate();
        if(result==1) {
            ResultSet keys=pst.getGeneratedKeys();
            keys.next();
            id=keys.getInt(1);
        }
        pst.close();
        return id;
    }
    @Override
    public boolean updateContract(ContractDB c) throws SQLException {
        boolean result=false;
        Connection conn=DBTool.getConnection();
        String sql="update contract set houseId=?,tenantId=?,salesmanId=?"
                + ", startDate=?, endDate=?,freOfPay=?, rentOfMonth=?,deposit=?"
                + ",signDate=?, state=?,remark=? where id=? ";
        SimpleDateFormat sdf=new SimpleDateFormat("yyyy-MM-dd");
        String signDate=sdf.format(new Date());
        PreparedStatement pst=conn.prepareStatement(sql);
        pst.setInt(1, c.getHouseId());
        pst.setString(2, c.getTenantId());
        pst.setInt(3, c.getSalesmanId());
        pst.setString(4, c.getStartDate());
        pst.setString(5, c.getEndDate());
        pst.setString(6, c.getFreOfPay());
        pst.setFloat(7, c.getRentOfMonth());
        pst.setFloat(8, c.getDeposit());
        pst.setString(9, signDate);
        pst.setString(10, c.getState());
        pst.setString(11, c.getRemark());
        pst.setInt(12, c.getId());
        int count=pst.executeUpdate();
        if(count==1) {
            result=true;
        }
        pst.close();
        return result;
```

```java
    }
    @Override
    public List<ContractInformation> searchAllContracts() {
        return DBTool.executeQuery(ContractInformation.class,sql,null);
    }
    @Override
    public List<ContractInformation> searchContracts(QueryCriteria criteria,
                                                     List<Object> opt) {
        String searchSql=sql+" where "+criteria.getCriteria();
        return DBTool.executeQuery(ContractInformation.class,searchSql,opt);
    }
}
```

5. 接口 IntentionDao 及其实现类 IntentionDaoImpl

```java
public interface IntentionDao {
    List<IntentionInformation> searchAllIntentionsByTenantId(String tenantId);
    int addIntention(IntentionInformation i) throws SQLException;
    boolean updateIntention(IntentionInformation i) throws SQLException;
    public boolean deleteIntentionById(String id) throws SQLException;
}
public class IntentionDaoImpl implements IntentionDao {
    @Override
    public int addIntention(IntentionInformation i) throws SQLException {
        int id=-1;
        Connection conn=DBTool.getConnection();
        String sql="insert into intention(tenantId,layout,area,decorate,
        price,remark) values (?,?,?,?,?,?)";
        PreparedStatement pst=conn.prepareStatement (sql,
                                        Statement.RETURN_GENERATED_KEYS);
        pst.setString(1, i.getTenantId());
        pst.setString(2, i.getLayout());
        pst.setFloat(3, i.getArea());
        pst.setString(4, i.getDecorate());
        pst.setFloat(5, i.getPrice());
        pst.setString(6, i.getRemark());
        int result=pst.executeUpdate();
        if(result==1) {
            ResultSet keys=pst.getGeneratedKeys();
            keys.next();
            id=keys.getInt(1);
        }
        pst.close();
        return id;
    }
```

```java
@Override
public boolean updateIntention(IntentionInformation i) throws SQLException {
    boolean result=false;
    Connection conn=DBTool.getConnection();
    String sql="update intention set layout=?,area=?,decorate=?,"
                                    +"price=?,remark=? where id=?";
    PreparedStatement pst=conn.prepareStatement(sql);
    pst.setString(1, i.getLayout());
    pst.setFloat(2, i.getArea());
    pst.setString(3, i.getDecorate());
    pst.setFloat(4, i.getPrice());
    pst.setString(5, i.getRemark());
    pst.setInt(6, i.getId());
    int count=pst.executeUpdate();
    if(count==1) {
        result=true;
    }
    pst.close();
    return result;
}
@Override
public boolean deleteIntentionById(String strId) throws SQLException {
    int id=Integer.parseInt(strId);
    boolean result=false;
    Connection conn=DBTool.getConnection();
    String sql="delete from intention where id=?";
    PreparedStatement pst=conn.prepareStatement(sql);
    pst.setInt(1, id);
    int count=pst.executeUpdate();
    if(count==1) {
        result=true;
    }
    pst.close();
    return result;
}
@Override
public List<IntentionInformation> searchAllIntentionsByTenantId(String
                                                            tenantId) {
    String sql="select id,layout,area,decorate,price,remark from intention"
                                        +" where tenantId="+tenantId;
    return DBTool.executeQuery(IntentionInformation.class, sql, null);
}
}
```

6. 接口 SalesmanDao 及其实现类 SalesmanDaoImpl

```java
public interface SalesmanDao {
    int addSalesman(SalesmanInformation s) throws SQLException;
    boolean updateSalesman(SalesmanInformation s) throws SQLException;
}
public class SalesmanDaoImpl implements SalesmanDao {
    @Override
    public int addSalesman(SalesmanInformation s) throws SQLException {
        int id=-1;
        Connection conn=DBTool.getConnection();
        String sql="insert into salesman(name,personalId,telephone,state)
        values(?,?,?,?)";
        PreparedStatement pst=conn.prepareStatement(sql,
                                        Statement.RETURN_GENERATED_KEYS);
        pst.setString(1, s.getName());
        pst.setString(2, s.getPersonalId());
        pst.setString(3, s.getTelephone());
        pst.setString(4, s.getState());
        int result=pst.executeUpdate();
        if(result==1) {
            ResultSet keys=pst.getGeneratedKeys();
            keys.next();
            id=keys.getInt(1);
        }
        pst.close();
        return id;
    }
    @Override
    public boolean updateSalesman(SalesmanInformation s) throws SQLException {
        boolean result=false;
        Connection conn=DBTool.getConnection();
        String sql="update salesman set name=?,personalId=?,telephone=?,
                                        state=? where id=?";
        PreparedStatement pst=conn.prepareStatement(sql);
        pst.setString(1, s.getName());
        pst.setString(2, s.getPersonalId());
        pst.setString(3, s.getTelephone());
        pst.setString(4, s.getState());
        pst.setInt(5, s.getId());
        int count=pst.executeUpdate();
        if(count==1) {
            result=true;
        }
```

```
        pst.close();
        return result;
    }
}
```

15.3.4 业务逻辑包 service

15.3.4

1. 接口 UserService 及其实现类 UserServiceImpl

```
public interface UserService {
    UserInformation login(int id, String pwd);
}
public class UserServiceImpl implements UserService {
    private UserDao dao;
    public UserServiceImpl( UserDao dao) {
        this.dao=dao;
    }
    @Override
    public UserInformation login(int id, String pwd) {
    return dao.login(id, pwd);
    }
}
```

2. 接口 HouseService 及其实现类 HouseServiceImpl

```
public interface HouseService {
    List<HouseInformation>searchAllHouses();
    List<HouseInformation>searchHouses(QueryCriteria criteria, List<Object>
                                                                opt);
    boolean addHouse(HouseDB h);
    boolean updateHouse(HouseDB h);
    HouseInformation searchHouseById(String id);
    boolean deleteHouseById(String id);
    boolean updateHouseStateById(int id, String state);
}
public class HouseServiceImpl implements HouseService {
    private HouseDao dao;
    public HouseServiceImpl(HouseDao dao) {
        this.dao=dao;
    }
    @Override
    public List<HouseInformation>searchAllHouses() {
        return dao.searchAllHouses();
    }
```

```java
@Override
public boolean addHouse(HouseDB h) {
    int id=-1;
    try {
        id=dao.addHouse(h);
    } catch (SQLException e) {
        e.printStackTrace();
    }
    if(id==-1)
        return false;
    else {
        h.setId(id);
        return true;
    }
}
@Override
public boolean updateHouse(HouseDB h) {
    try {
        return dao.updateHouse(h);
    } catch (SQLException e) {
        e.printStackTrace();
        return false;
    }
}
@Override
public boolean deleteHouseById(String id) {
    return dao.deleteHouseById(id);
}
@Override
public List<HouseInformation> searchHouses(QueryCriteria criteria, List<Object>opt) {
    return dao.searchHouses(criteria, opt);
}
@Override
public HouseInformation searchHouseById(String id) {
    return dao.searchHouseById(id);
}
@Override
public boolean updateHouseStateById(int id, String state) {
    HouseInformation house=dao.searchHouseById(id+"");
    house.setState(state);
    try {
        return dao.updateHouse(new HouseDB(house));
    } catch (SQLException e) {
```

```java
            e.printStackTrace();
            return false;
        }
    }
}
```

3. 接口 CustomService 及其实现类 CustomServiceImpl

```java
public interface CustomService {
    CustomInformation searchCustomByPersonalId(String pid);
    List<CustomInformation> searchAllCustoms();
    List<CustomInformation> searchCustoms(QueryCriteria criteria, List
                                            <Object>opt);
    boolean addCustom(CustomInformation c);
    boolean updateCustom(CustomInformation t);
}
public class CustomServiceImpl implements CustomService {
    private CustomDao dao;
    public CustomServiceImpl(CustomDao dao) {
        this.dao=dao;
    }
    @Override
    public CustomInformation searchCustomByPersonalId(String pid) {
        return dao.searchCustomByPersonalID(pid);
    }
    @Override
    public List<CustomInformation> searchAllCustoms() {
        return dao.searchAllCustoms();
    }
    @Override
    public List<CustomInformation> searchCustoms(QueryCriteria criteria, List
                                            <Object>opt){
        return dao.searchCustoms(criteria, opt);
    }
    @Override
    public boolean addCustom(CustomInformation c) {
        try {
            return dao.addCustom(c);
        } catch (SQLException e) {
            e.printStackTrace();
            return false;
        }
    }
    @Override
    public boolean updateCustom(CustomInformation t) {
```

```java
        try {
            return dao.updateCustom(t);
        } catch (SQLException e) {
            e.printStackTrace();
            return false;
        }
    }
}
```

4. 接口 ContractService 及其实现类 ContractServiceImpl

```java
public interface ContractService {
    List<ContractInformation> searchAllContracts();
    List<ContractInformation> searchContracts(QueryCriteria criteria, List
                                                        <Object> opt);
    boolean addContract(ContractDB c) throws Exception;
    boolean teminateContractById(ContractInformation c);
}
public class ContractServiceImpl implements ContractService {
    private ContractDao contractDao;
    private HouseService houseServ;
    public ContractServiceImpl( ContractDao dao, HouseService houseServ) {
        this.contractDao=dao;
        this.houseServ=houseServ;
    }
    @Override
    public List<ContractInformation> searchAllContracts() {
        return contractDao.searchAllContracts();
    }
    @Override
    public List<ContractInformation> searchContracts(QueryCriteria criteria,
                                                        List<Object> opt) {
        return contractDao.searchContracts(criteria, opt);
    }
    @Override
    public boolean addContract(ContractDB c) throws Exception{
        //判断房屋状态是否为出租
        HouseInformation house=houseServ.searchHouseById(c.getHouseId()+"");
        if(house.getState().equals(Constants.STATE_HOUSE[2])) {
            throw new Exception("房屋已经出租.");
        }
        //添加合同
        int id=-1;
        try {
            id=contractDao.addContract(c);
        } catch (SQLException e) {
```

```java
            e.printStackTrace();
        }
        if(id==-1)
            return false;
        else {
            //修改房屋状态
            c.setId(id);
            house.setState(Constants.STATE_HOUSE[2]);
            return houseServ.updateHouseStateById(house.getId()
                                    , Constants.STATE_HOUSE[2]);
        }
    }
    @Override
    public boolean teminateContractById(ContractInformation c) {
        ContractDB dbC=new ContractDB(c);
        dbC.setState(Constants.STATE_CONTRACT[2]);
        int houseId=dbC.getHouseId();
        try {
            if( contractDao.updateContract(dbC)) {
                //修改房屋状态
                return houseServ.updateHouseStateById(houseId
                                    , Constants.STATE_HOUSE[1]);
            }
            else
                return false;
        } catch (SQLException e) {
            e.printStackTrace();
            return false;
        }
    }
}
```

5. 接口 IntentionService 及其实现类 IntentionServiceImpl

```java
public interface IntentionService {
    List<IntentionInformation> searchAllIntentionsByTenantId(String tenantId);
    boolean addIntention(IntentionInformation i);
    boolean updateIntention(IntentionInformation i);
    boolean deleteIntentionById(String id);
}
public class IntentionServiceImpl implements IntentionService {
    private IntentionDao dao;
    public IntentionServiceImpl(IntentionDao dao) {
        this.dao=dao;
    }
    @Override
```

```java
        public List<IntentionInformation> searchAllIntentionsByTenantId(String
                                                            tenantId) {
            return dao.searchAllIntentionsByTenantId(tenantId);
        }
        @Override
        public boolean addIntention(IntentionInformation i) {
            int id=-1;
            try {
                id=dao.addIntention(i);
            } catch (SQLException e) {
                e.printStackTrace();
            }
            if(id==-1)
                return false;
            else {
                i.setId(id);
                return true;
            }
        }
        @Override
        public boolean updateIntention(IntentionInformation i) {
            try {
                return dao.updateIntention(i);
            } catch (SQLException e) {
                e.printStackTrace();
                return false;
            }
        }
        @Override
        public boolean deleteIntentionById(String id) {
            try {
                return dao.deleteIntentionById(id);
            } catch (SQLException e) {
                e.printStackTrace();
                return false;
            }
        }
    }
```

6. 接口 SalesmanService 及其实现类 SalesmanServiceImpl

```java
public interface SalesmanService {
    SalesmanInformation searchSalesmanById(String id);
    List<SalesmanInformation> searchAllSalesman();
    List<SalesmanInformation> searchSalesman(QueryCriteria criteria, List
```

```java
                                            <Object>opt);
    boolean addSalesman(SalesmanInformation s);
    boolean deleteSalesmanById(String id);
    boolean updateSalesman(SalesmanInformation s);
}
public class SalesmanServiceImpl implements SalesmanService {
    private SalesmanDao dao;
    public SalesmanServiceImpl(SalesmanDao dao) {
        this.dao=dao;
    }
    @Override
    public SalesmanInformation searchSalesmanById(String id) {
        String sql="select * from salesman where id=? ";
        List<Object>params=new ArrayList<Object>();
        params.add(id);
        List<SalesmanInformation>list=DBTool.executeQuery(
                            SalesmanInformation.class, sql, params);
        if(list !=null && list.size()>0)
            return list.get(0);
        else
            return null;
    }
    @Override
    public List<SalesmanInformation>searchAllSalesman() {
        String sql="select * from salesman ";
        return DBTool.executeQuery(SalesmanInformation.class, sql, null);
    }
    @Override
    public List< SalesmanInformation > searchSalesman(QueryCriteria criteria,
                                            List<Object>opt) {
        String sql="select * from salesman where "+criteria.getCriteria();
        return DBTool.executeQuery(SalesmanInformation.class, sql, opt);
    }
    @Override
    public boolean addSalesman(SalesmanInformation s) {
        int id=-1;
        try {
            id=dao.addSalesman(s);
        } catch (SQLException e) {
            e.printStackTrace();
        }
        if(id==-1)
            return false;
        else {
```

```java
            s.setId(id);
            return true;
        }
    }
    @Override
    public boolean deleteSalesmanById(String id) {
        boolean result=false;
        String sql="select * from salesman where id=?";
        List<Object>params=new ArrayList<Object>();
        params.add(id);
        List<SalesmanInformation>list=DBTool.executeQuery(
                        SalesmanInformation.class, sql, params);
        if(list !=null) {
            SalesmanInformation s=list.get(0);
            s.setState("离职");
            try {
                result=dao.updateSalesman(s);
            } catch (SQLException e) {
                e.printStackTrace();
            }
        }
        return result;
    }
    @Override
    public boolean updateSalesman(SalesmanInformation s) {
        try {
            return dao.updateSalesman(s);
        } catch (SQLException e) {
            e.printStackTrace();
            return false;
        }
    }
}
```

15.3.5 工具包 util

1. 数据库工具类 DBTool

DBTool 封装与远程数据库的连接对象,并对外提供 3 个公有方法。

```java
public class DBTool {
    private static Connection conn =null;
    //获得数据库的连接对象
    public static Connection getConnection(){
```

```java
        try {
            if(conn==null || conn.isClosed()){
                try {
                    Class.forName("com.mysql.cj.jdbc.Driver");
                    conn = DriverManager.getConnection("jdbc:mysql://
                        localhost:3306/house?useSSL=false&serverTimezone=
                        UTC","yulp","123456");
                } catch (ClassNotFoundException e) {
                    e.printStackTrace();
                } catch (SQLException e) {
                    e.printStackTrace();
                }
            }
        } catch (SQLException e) {
            e.printStackTrace();
        }
        return conn;
    }
    //关闭数据库的连接
    public static void closeConnection(){
        try {
            if(conn!=null && !conn.isClosed()){
                conn.close();
                conn=null;
            }
        }catch (SQLException e) {
            e.printStackTrace();
        }
    }
    /**
     * 通用查询方法
     * @param clazz 查询获取的对象类型
     * @param sql 包含 select 的 SQL 语句
     * @param args 与 SQL 语句中"?"相对应的参数值
     * @return 查询得到满足条件的对象集合
     */
    public static <T>List<T>executeQuery(Class<T>clazz, String sql,
                                        List<Object>args) {
        Connection conn=null;
        PreparedStatement preparedstatement=null;
        ResultSet rs=null;
        List<T>vecRs=new ArrayList<T>();
        T obj=null;
        try {
```

```java
            conn=getConnection();
            preparedstatement=conn.prepareStatement(sql);
            if (args!=null && args.size()!=0)
                for (int i=0; i<args.size(); i++)
                    preparedstatement.setObject(i+1, args.get(i));
            rs=preparedstatement.executeQuery();
            //获取元数据
            ResultSetMetaData rsmd=rs.getMetaData();
            Map<String, Object>mapMetaData=new HashMap<String, Object>();
            //打印一列的列名
            while (rs.next()) {
                //获取数据表中满足要求的一行数据,并放入 Map 中
                for (int i=0; i<rsmd.getColumnCount(); i++) {
                    String columnLabel=rsmd.getColumnLabel(i+1);
                    Object columnValue=rs.getObject(columnLabel);
                    System.out.println(columnLabel);
                    mapMetaData.put(columnLabel, columnValue);
                }
                //将 Map 中的数据通过反射初始化 T 类型对象
                if (mapMetaData.size()>0) {
                    obj=clazz.newInstance();
                    for(Map.Entry<String, Object>entry: mapMetaData.entrySet())
                    {
                        String fieldkey=entry.getKey();
                        Object fieldvalue=entry.getValue();
                        System.out.println(fieldkey+":"+fieldvalue);
                        setFieldValue(obj, fieldkey, fieldvalue);
                    }
                }
                vecRs.add(obj);        //将对象装入 Vector 容器
            }
        } catch (Exception e) {
            e.printStackTrace();
        }
        return vecRs;
    }
    //设置对象的属性
    private static void setFieldValue(Object obj, String fieldName, Object
                                                                    value) {
        Field field=getDeclaredField(obj, fieldName);
        if (field==null) {
            throw new IllegalArgumentException("Could not find field["
                                +fieldName+"] on target ["+obj+"]");
        }
```

```java
        makeAccessiable(field);
        try {
            if (value !=null)
                field.set(obj, value);
        } catch (Exception e) {
            e. printStackTrace();
        }
    }
    //判断 field 的修饰符是否是 public,并据此改变 field 的访问权限
    private static void makeAccessiable(Field field) {
        if (!Modifier.isPublic(field.getModifiers())) {
            field.setAccessible(true);
        }
    }
    //获取 field 属性,属性有可能在父类中继承
    private static Field getDeclaredField(Object obj, String fieldName) {
        for (Class<?>clazz=obj.getClass(); clazz !=Object.class
                                    ; clazz=clazz.getSuperclass()) {
            try {
                return clazz.getDeclaredField(fieldName);
            } catch (Exception e) {
            }
        }
        return null;
    }
}
```

2. 常量定义接口 Constants

接口 Constants 定义项目中使用的常量。

```java
public interface Constants {
    String[] STYLE_DECORATING={"","精装修","简装修","毛坯房"};
    String[] FREQUENCY_PAY={"","年缴","季缴","月缴"};
    String[] STATE_HOUSE=new String[]{"","待租","出租"};
    String[] STATE_CONTRACT=new String[]{"","有效","提前解约","结束"};
    String[] STATE_SALESMAN=new String[] {"","在职","离职"};
    Operator[] OPERATORS={new Operator("等于","=")
        ,new Operator("小于","<"),new Operator("小于或等于","<=")
        ,new Operator("大于",">"),new Operator("大于或等于",">=")
        ,new Operator("包含","like") };
}
```

15.3.6 图形用户界面包 gui

1. 登录

登录功能的效果如图 15-19 所示。

图 15-19 登录界面

省略界面布局的部分代码，核心代码如下。

```java
package gui;
public class LoginGUI extends JFrame {
    private JTextField jtf_id;
    private JPasswordField jtf_password;
    private JButton btnClear;
    private JButton btnLogin;
    public LoginGUI () {
        super("登录");
        jtf_id=new JTextField(8);
        jtf_password=new JPasswordField(8);
        JLabel lbl_id=new JLabel("id:");
        JLabel lbl_password=new JLabel("密码:");
        btnLogin=new JButton("登录");
        btnLogin.addActionListener(new ActionListener() {
            @Override
            public void actionPerformed(ActionEvent e) {
                String strId=jtf_id.getText();
                String pwd=new String(jtf_password.getPassword());
                int id=-1;
                try {
                    id=Integer.parseInt(strId);
```

```java
            } catch (NumberFormatException ee) {
                JOptionPane.showMessageDialog(null, "ID必须为数字",
                                null, JOptionPane.ERROR_MESSAGE);
                return;
            }
            UserService service=new UserServiceImpl(new UserDaoImpl());
            User m=service.login(id, pwd);
            if (m==null) {
                JOptionPane.showMessageDialog(null, "账号错误",
                                null, JOptionPane.ERROR_MESSAGE);
                return;
            }
            if (!m.getPassword().equals(pwd)) {
                JOptionPane.showMessageDialog(null, "密码错误",
                                null, JOptionPane.ERROR_MESSAGE);
                return;
            }
            new MainGUI();           //显示主界面
            Login.this.dispose();
        }
    });
    btnClear=new JButton("重新填写");
    btnClear.addActionListener(new ActionListener() {
        @Override
        public void actionPerformed(ActionEvent e) {
            jtf_id.setText("");
            jtf_password.setText("");
        }
    });
}
public static void main(String[] args) {
    new LoginGUI ();
}
}
```

2. 主界面

主界面主要由选项卡面板JTabbedPane构成，效果如图15-20所示。若干个选项卡面板都继承自抽象父类RefreshPanel，其定义如下。

```java
public abstract class RefreshPanel extends JPanel{
    public abstract void refresh();
}
```

每个选项卡子类都需要重写refresh()方法，以实现数据的刷新操作。当用户单击选

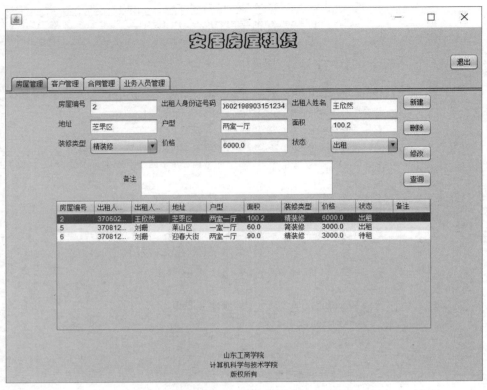

图 15-20 主界面

项卡标题实现选项卡切换时,通过 ChangeListener 监听器调用被选中选项卡的 refresh()方法,实现数据的实时更新。

核心代码如下。

```java
public class MainGUI extends JFrame {
    private JButton btnLogout;
    public MainGUI() {
        Container c=this.getContentPane();
        JPanel palTop=new JPanel(new BorderLayout());
        JPanel titlePal=new JPanel();
        JLabel title=new JLabel("安居房屋租赁");
        title.setFont(new Font("华文彩云",Font.BOLD,30));
        titlePal.add(title);
        palTop.add(titlePal);
        btnLogout=new JButton("退出");
        btnLogout.addActionListener(new ActionListener() {
            @Override
            public void actionPerformed(ActionEvent arg0) {
                System.exit(0);
            }
```

```java
        });
        JPanel topRight=new JPanel(new FlowLayout(FlowLayout.RIGHT));
        topRight.add(btnLogout);
        palTop.add(topRight,BorderLayout.SOUTH);
        JTabbedPane center=new JTabbedPane(JTabbedPane.TOP);
        RefreshPanel house=new HouseManageGUI();
        RefreshPanel custom=new CustomManageGUI();
        RefreshPanel contract=new ContractPal();
        RefreshPanel salesman=new SalesmanManageGUI();
        center.add("房屋管理", house);
        center.add("客户管理", custom);
        center.add("合同管理", contract);
        center.add("业务人员管理", salesman);
        center.addChangeListener(new ChangeListener(){
            @Override
            public void stateChanged(ChangeEvent e) {
                JTabbedPane tab=(JTabbedPane)(e.getSource());
                ((RefreshPanel)tab.getSelectedComponent()).refresh();
            }
        });
        Box bottom=Box.createVerticalBox();
        JLabel lbl1=new JLabel("山东工商学院");
        lbl1.setAlignmentX(CENTER_ALIGNMENT);
        JLabel lbl2=new JLabel("计算机科学与技术学院");
        lbl2.setAlignmentX(CENTER_ALIGNMENT);
        JLabel lbl3=new JLabel("版权所有");
        lbl3.setAlignmentX(CENTER_ALIGNMENT);
        bottom.add(lbl1);
        bottom.add(lbl2);
        bottom.add(lbl3);
        JPanel palBottom=new JPanel();
        palBottom.add(bottom);
        c.add(palTop, BorderLayout.NORTH);
        c.add(center);
        c.add(palBottom,BorderLayout.SOUTH);
        this.setBounds(100,200,800,600);
        this.setDefaultCloseOperation(JFrame.EXIT_ON_CLOSE);
        this.setVisible(true);
    }
}
```

3. 主信息面板

根据需求规格说明，系统包含房屋管理、客户管理、合同管理、意向管理、业务员管理等主要人机交互界面,界面具有相似结构。通过对上述界面进行抽象分析,得到以下主信息面板类 InformationPal,效果如图 15-21 所示。

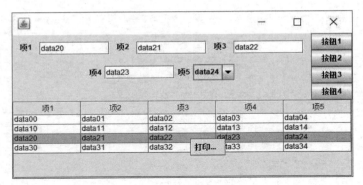

图 15-21　主信息面板

InformationPal 类中封装如下数据。

```
private List<JLabel>lblItemNames;
private List<JComponent>cptItems;
private List<JButton>btn;
private JTable tblInfo;
private DefaultTableModel model;
```

其中,lblItemNames 定义界面中包含的标签元素;cptItems 定义输入或者选择组件;btn 定义界面右侧的按钮组件列表;tblInfo 和 model 分别用于定义主信息面板下方的表格组件和表格模型。

为表格组件 tblInfo 添加鼠标事件监听器,单击表格 tblInfo,将当前行的内容显示在主信息面板上方的编辑或选择组件中。

为表格组件 tblInfo 添加快捷菜单,实现报表打印功能。

核心代码如下。

```
public class InformationPal extends JPanel{
/**
 * @param itemNames: 表格 tblInfo 的列名称或者信息面板上方的信息明细
 * @param items: 信息面板上方的输入或选择组件
 * @param btnCommond: 信息面板右侧的按钮列表
 * @param countOneLine: 信息面板上方每行显示的信息明细项数
 * @throws Exception
 */
public InformationPal (List<String>itemNames,List<JComponent>items
        ,List<JButton>btnCommond, int countOneLine) throws Exception{
    tableColNames=new Vector<String>(itemNames);
    lblItemNames=new ArrayList<JLabel>();
    for(int i=0; i<itemNames.size(); i++)
        lblItemNames.add(new JLabel(itemNames.get(i)));
```

```java
cptItems=items;
btn=btnCommond;
palDetail=new JPanel(new BorderLayout());
Box commond=Box.createVerticalBox();
for(JButton b:btn) {
    commond.add(b);
    commond.add(Box.createGlue());
}
palDetail.add(commond,BorderLayout.EAST);
JPanel topCenter=new JPanel();
GroupLayout layout=new GroupLayout(topCenter);
topCenter.setLayout(layout);
layout.setAutoCreateContainerGaps(true);
layout.setAutoCreateGaps(true);
GroupLayout.ParallelGroup groupLbl[]=new GroupLayout
                                    .ParallelGroup[countOneLine];
GroupLayout.ParallelGroup[] groupComponent=new GroupLayout
                                    .ParallelGroup[countOneLine];
GroupLayout.SequentialGroup groupH=layout
                                    .createSequentialGroup();
for(int i=0; i<countOneLine; i++) {
    groupLbl[i]=layout.createParallelGroup(
                            GroupLayout.Alignment.LEADING);
    groupComponent[i]=layout.createParallelGroup(
                            GroupLayout.Alignment.LEADING);
    for(int j=i; j<countOneLine*(lblItemNames.size()/countOneLine)
                                    ; j+=countOneLine) {
        groupLbl[i].addComponent(lblItemNames.get(j));
        groupComponent[i].addComponent(cptItems.get(j));
    }
    groupH.addGroup(groupLbl[i]).addGroup(groupComponent[i]);
}
layout.setHorizontalGroup(groupH);
GroupLayout.ParallelGroup groupline[]=new GroupLayout
            .ParallelGroup[lblItemNames.size()/countOneLine];
GroupLayout.SequentialGroup groupV=layout.createSequentialGroup();
int currentComp=0;
for(int i=0; i<groupline.length; i++) {
    groupline[i]=layout.createParallelGroup(
                        GroupLayout.Alignment.LEADING);
    for(int count=0; count<countOneLine; count++) {
        groupline[i].addComponent(lblItemNames.get(currentComp));
        groupline[i].addComponent(cptItems.get(currentComp++));
    }
```

```java
            groupV.addGroup(groupline[i]);
    }
    layout.setVerticalGroup(groupV);
    JPanel residue=new JPanel();
    for(int k=countOneLine * (lblItemNames.size()/countOneLine);
                                k<lblItemNames.size(); k++) {
        residue.add(lblItemNames.get(k));
        residue.add(cptItems.get(k));
    }
    Box g=Box.createVerticalBox();
    g.add(topCenter);
    g.add(residue);
    palDetail.add(g);
    model=new DefaultTableModel(null,tableColNames);
    tblInfo=new JTable(model);
    tblInfo.addMouseListener(new MouseAdapter() {
        @Override
        public void mouseClicked(MouseEvent arg0) {
            int currentRow=tblInfo.getSelectedRow();
            if(currentRow==-1)
                return ;
            for(int i=0; i<cptItems.size();i++) {
                JComponent c=cptItems.get(i);
                String v=(String)tblInfo.getValueAt(currentRow, i);
                if(v==null)
                    v="";
                if(c instanceof JTextComponent)
                    ((JTextComponent)c).setText(v);
                else if(c instanceof JComboBox)
                    ((JComboBox)c).setSelectedItem(v);
                else if(c instanceof JScrollPane) {
                    @SuppressWarnings("unchecked")
                    JList<String>list=(JList<String>)
                        (((JScrollPane)c).getViewport().getView());
                    DefaultListModel<String>listModel=
                        (DefaultListModel<String>)(list.getModel());
                    String[] value=v.split(";");
                    listModel.removeAllElements();
                    for (String s : value) {
                        listModel.addElement(s);
                    }
                }
            }
        }
```

```java
            @Override
            public void mousePressed(MouseEvent arg0) {
                if(arg0.isPopupTrigger())
                    popUpMenu.show(tblInfo, arg0.getX(), arg0.getY());
            }
        });
        setLayout(new BorderLayout());
        add(palDetail,BorderLayout.NORTH);
        add(new JScrollPane(tblInfo));
        popUpMenu=new JPopupMenu();
        JMenuItem printMenu=new JMenuItem("打印...");
        popUpMenu.add(printMenu);
        printMenu.addActionListener(new ActionListener() {
            @Override
            public void actionPerformed(ActionEvent arg0) {
                try {
                    tblInfo.print();
                } catch (PrinterException e) {
                    e.printStackTrace();
                }
            }
        });
    }
    public void setDetailPanelSize(Dimension d) {
        palDetail.setSize(d);
    }
    public void freshTable(Vector<Vector<String>>data) {
        model.setDataVector(data, tableColNames);
        model.fireTableDataChanged();
    }
    public void freshTable(List<? >data) {
        Vector<Vector<String>>vec=new Vector<Vector<String>>();
        for(Object o:data) {
            String[] sDetails=o.toString().split(",");
            Vector<String>line=new Vector<String>();
            for(String str:sDetails)
                line.add(str);
            vec.add(line);
        }
        freshTable(vec);
    }
}
```

4. 房屋管理面板

房屋管理面板 HouseManageGUI 位于主界面的首页,用于实现房屋的新建、删除、修改与查询操作。

HouseManageGUI 类中封装了以下数据。

```java
private InformationPal houseInfo;
private List<String>lblItemNames;
private List<JComponent>cptItems;
private List<JButton>btnCommond;
private Custom owner;
private HouseService houseServ;
private CustomService customServ;
```

其中，主信息面板对象 houseInfo 用于显示及输入房屋信息，lblItemNames、cptItems 及 btnCommond 用于构造对象 houseInfo，具体参见主信息面板的说明，owner 对象用于记录当前房屋的房主信息，HouseManageGUI 中封装业务服务对象 houseServ 和 customServ，分别提供对房屋及房屋出租人的操作。

核心代码如下。

```java
public class HouseManageGUI extends RefreshPanel implements ActionListener {
    public HouseManageGUI() {
        lblItemNames=new ArrayList<String>();
        lblItemNames.add("房屋编号");
        lblItemNames.add("出租人身份证号码");
        lblItemNames.add("出租人姓名");
        lblItemNames.add("地址");
        lblItemNames.add("户型");
        lblItemNames.add("面积");
        lblItemNames.add("装修类型");
        lblItemNames.add("价格");
        lblItemNames.add("状态");
        lblItemNames.add("备注");
        cptItems=new ArrayList<JComponent>();
        for(int i=0; i<6;i++) {
            JTextField t=new JTextField(10);
            t.setMaximumSize(t.getPreferredSize());
            cptItems.add(t);
        }
        ((JTextField)cptItems.get(0)).setEditable(false);
        JTextField ownerId=(JTextField)cptItems.get(1);
            //根据房主身份证查询房主
        ownerId.addActionListener(new ActionListener() {
            @Override
            public void actionPerformed(ActionEvent e) {
                owner=customServ.searchCustomByPersonalId(ownerId.getText());
                if(owner==null) {
```

```java
                    JOptionPane.showMessageDialog(HouseManageGUI.this
                            , "房主信息不存在,请先完善房主信息");
                    return;
                }
                ((JTextField)cptItems.get(2)).setText(owner.getName());
            }
        });
        JComboBox com1=new JComboBox(Constants.STYLE_DECORATING);
        com1.setPreferredSize(cptItems.get(0).getPreferredSize());
        com1.setMaximumSize(com1.getPreferredSize());
        cptItems.add(com1);
        JTextField t=new JTextField(10);
        t.setMaximumSize(t.getPreferredSize());
        cptItems.add(t);
        JComboBox com2=new JComboBox(Constants.STATE_HOUSE);
        com2.setPreferredSize(cptItems.get(0).getPreferredSize());
        com2.setMaximumSize(com2.getPreferredSize());
        cptItems.add(com2);
        JTextArea remark=new JTextArea(3,30);
        cptItems.add(remark);
        btnCommond=new ArrayList<JButton>();
        btnCommond.add(new JButton("新建"));
        btnCommond.add(new JButton("删除"));
        btnCommond.add(new JButton("修改"));
        btnCommond.add(new JButton("查询"));
        for(int i=0; i<btnCommond.size();i++) {
            btnCommond.get(i).addActionListener(this);
        }
        try {
            houseInfo=new InformationPal(lblItemNames,cptItems,btnCommond,3);
        } catch (Exception e) {
            e.printStackTrace();
        }
        houseServ=new HouseServiceImpl(new HouseDaoImpl());
        customServ=new CustomServiceImpl(new CustomDaoImpl());
        refresh();
        houseInfo.setDetailPanelSize(new Dimension(400,500));
        add(houseInfo);
    }
    public void refresh(){
        List<House>houses=houseServ.searchAllHouses();
        houseInfo.freshTable(houses);
    }
    public void clear() {
```

```java
            for(int i=0; i<6; i++)
                ((JTextField)cptItems.get(i)).setText("");
            ((JComboBox)cptItems.get(6)).setSelectedItem("");
            ((JTextField)cptItems.get(7)).setText("");
            ((JComboBox)cptItems.get(8)).setSelectedItem("");
    }
    private House getInputHouseInfo() {
        House h=null;
        String strOwnerId=((JTextField)cptItems.get(1)).getText();
        String strOwnerName=((JTextField)cptItems.get(2)).getText();
        if(owner==null)
            owner=customServ.searchCustomByPersonalId(strOwnerId);
        if(owner==null || !(owner.getName()).equals(strOwnerName)) {
            JOptionPane.showMessageDialog(HouseManageGUI.this
                            ,"房主信息不正确,请确认房主信息.");
        }else {
            int id=0;
            String strId=((JTextField)cptItems.get(0)).getText();
            if(!strId.equals(""))
                id=Integer.parseInt(strId);
            String address=((JTextField)cptItems.get(3)).getText();
            String layout=((JTextField)cptItems.get(4)).getText();
            float area=Float.parseFloat(((JTextField)cptItems.get(5)).getText());
            String decoration=(String)(((JComboBox)cptItems.get(6))
            .getSelectedItem());
            float price=Float.parseFloat(((JTextField)cptItems.get(7)).getText());
            String state=(String)(((JComboBox)cptItems.get(8)).getSelectedItem());
            String remark=(String)(((JTextArea)cptItems.get(9)).getText());
            h=new House(id,owner.getPersonalId(),owner.getName(),address
                            ,layout,area,decoration,price,state,remark);
        }
        owner=null;
        return h;
    }
    @Override
    public void actionPerformed(ActionEvent e) {
        String strCmd=e.getActionCommand();
        JButton btn=(JButton)e.getSource();
        if("新建".equals(strCmd)){
            clear();
            btn.setText("保存");
        }else if("删除".equals(strCmd)) {
            String strHouseId=((JTextField)cptItems.get(0)).getText();
            if(strHouseId==null || strHouseId.equals("")) {
```

```java
            JOptionPane.showMessageDialog(HouseManageGUI.this
                                            ,"请选择待删除的房屋");
        return;
    }
    if(((JComboBox)cptItems.get(8)).getSelectedItem().equals("出租")) {
        JOptionPane.showMessageDialog(HouseManageGUI.this
                                        ,"房屋出租中,无法删除");
        return;
    }
    int confirm=JOptionPane.showConfirmDialog(HouseManageGUI.this
                                    ,"确认删除当前房屋信息吗?");
    if(confirm==JOptionPane.YES_OPTION) {
        if(houseServ.deleteHouseById(strHouseId)) {
            JOptionPane.showMessageDialog(HouseManageGUI.this
                                            ,"删除成功");
            clear();
            refresh();
        }else
            JOptionPane.showMessageDialog(HouseManageGUI.this
                                            ,"删除失败");
    }
}else if("保存".equals(strCmd)) {
    House inputHouse=getInputHouseInfo();
    if(inputHouse==null)
        return;
    btn.setText("新建");
    if(houseServ.addHouse(new HouseDB(inputHouse))) {
        JOptionPane.showMessageDialog(HouseManageGUI.this, "创建成功");
        ((JTextField)cptItems.get(0)).setText(""+inputHouse.getId());
        refresh();
    }else
        JOptionPane.showMessageDialog(HouseManageGUI.this, "创建失败");
}else if("查询".equals(strCmd)) {
    List<String>dbItemNames=new ArrayList<String>();
    dbItemNames.add("house.id");
    dbItemNames.add("custom.personalId");
    dbItemNames.add("custom.name");
    dbItemNames.add("house.address");
    dbItemNames.add("house.layout");
    dbItemNames.add("house.area");
    dbItemNames.add("house.decorate");
    dbItemNames.add("house.price");
    dbItemNames.add("house.state");
    dbItemNames.add("house.remark");
```

```java
            Vector<QueryColumn>columnNames=new Vector<QueryColumn>();
            for(int i=0; i<dbItemNames.size();i++)
                columnNames.add(new QueryColumn(lblItemNames.get(i),
                                                dbItemNames.get(i)));
            Vector<Operator>opers=new Vector<Operator>();
            for(int i=0; i<Constants.OPERATORS.length;i++)
                opers.add(Constants.OPERATORS[i]);
            QueryCriteriaInputGUI query=new QueryCriteriaInputGUI(
                                                columnNames,opers);
            query.setTitle("房屋查询");
            query.setBounds(HouseManageGUI.this.getX(), HouseManageGUI
                                                .this.getY(), 600, 400);
            query.setModal(true);
            query.setVisible(true);
            QueryCriteria criteria=query.getCriteria();
            List<Object>opt=query.getOpt();
            List<House>houses=houseServ.searchHouses(criteria, opt);
            if(houses==null || houses.size()==0)
                JOptionPane.showMessageDialog(HouseManageGUI.this
                                        , "没有满足条件的记录");
            houseInfo.freshTable(houses);
        }else if("修改".equals(strCmd)) {
            int confirm=JOptionPane.showConfirmDialog(HouseManageGUI.this
                                        , "确认修改当前房屋信息吗?");
            if(confirm==JOptionPane.YES_OPTION) {
                House inputHouse=getInputHouseInfo();
                if(houseServ.updateHouse(new HouseDB(inputHouse))) {
                    JOptionPane.showMessageDialog(HouseManageGUI.this
                                                , "修改成功");
                    refresh();
                }else
                    JOptionPane.showMessageDialog(HouseManageGUI.this
                                                , "修改失败");
            }
        }
    }
}
```

5. "房屋查询"面板

"房屋查询"面板用于获取查询条件,功能效果如图 15-22 所示。

(1) 查询面板子类 QueryCriteriaInputGUI。

项目涉及房屋查询、客户查询、合同查询及业务人员的查询操作,构造查询面板子类 QueryCriteriaInputGUI。

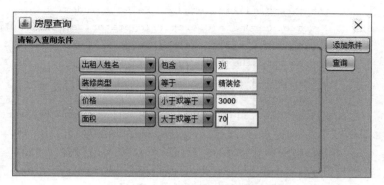

图 15-22 "房屋查询"面板

核心代码如下。

```java
public class QueryCriteriaInputGUI extends JDialog {
    private JComboBox<Column> columns;
    private JComboBox<Operator> operator;
    private JTextField value;
    private Vector<String> dbColumns;
    private QueryCriteria criteria;
    private List<Object> opt;
    public QueryCriteriaInputGUI(Vector<Column> columnNames, Vector
                                        <Operator> oper) {
        dbColumns=new Vector<String>();
        for (Column c : columnNames)
            dbColumns.add(c.getDbColumn());
        criteria=new QueryCriteria(dbColumns);
        columns=new JComboBox<Column>(columnNames);
        operator=new JComboBox<Operator>(oper);
        value=new JTextField("请输入数值");
        opt=new ArrayList<Object>();
        Container c=this.getContentPane();
        JPanel center=new JPanel();
        center.setBorder(BorderFactory.createTitledBorder("请输入查询条件"));
        Box boxLeft=Box.createVerticalBox();
        Box line=Box.createHorizontalBox();
        line.add(columns);
        line.add(operator);
        line.add(value);
        boxLeft.add(line);
        JButton btnAdd=new JButton("添加条件");
        btnAdd.addActionListener(new ActionListener() {
            @Override
            public void actionPerformed(ActionEvent arg0) {
```

```java
                    int columnIndex=columns.getSelectedIndex();
                    Operator curOper=(Operator)(operator.getSelectedItem());
                    String o=curOper.getValue();
                    String strValue=value.getText();
                    if(o.equals("like"))
                        strValue="% "+strValue+"% ";
                    if(strValue.equals("请输入数值")||strValue.equals(""))
                                                                return;
                    opt.add(strValue);
                    criteria.setColumnMatch(columnIndex, o);
                    Box l=Box.createHorizontalBox();
                    columns=new JComboBox<Column>(columnNames);
                    operator=new JComboBox<Operator>(oper);
                    value=new JTextField("请输入数值");
                    l.add(columns);
                    l.add(operator);
                    l.add(value);
                    boxLeft.add(l);
                    QueryCriteriaInputGUI.this.validate();
                    QueryCriteriaInputGUI.this.repaint();
                }
            });
            JButton btnQuery=new JButton("查询");
            btnQuery.addActionListener(new ActionListener() {
                @Override
                public void actionPerformed(ActionEvent arg0) {
                    int columnIndex=columns.getSelectedIndex();
                    String o=((Operator)(operator.getSelectedItem())).getValue();
                    String strValue=value.getText();
                    if(o.equals("like"))
                        strValue="% "+strValue+"% ";
                    if(!strValue.equals("请输入数值")&&!strValue.equals("")) {
                        opt.add(strValue);
                        criteria.setColumnMatch(columnIndex, o);
                    }
                    QueryCriteriaInputGUI.this.dispose();
                }
            });
            JPanel boxRight=new JPanel();
            boxRight.setLayout(new BoxLayout(boxRight,BoxLayout.Y_AXIS));
            boxRight.add(btnAdd);
            boxRight.add(btnQuery);
            center.add(boxLeft);
            c.add(center);
```

```
        c.add(boxRight, BorderLayout.EAST);
    }
    public QueryCriteria getCriteria() {
        return criteria;
    }
    public List<Object> getOpt() {
        return opt;
    }
}
```

（2）查询条件类 QueryCriteria。

columnNames 表示查询条件中的字段名，通过 getCriteria() 方法获取最终的查询条件。

```
public class QueryCriteria {
    private Vector<String> columnNames;
    private StringBuilder criteria=new StringBuilder();
    public String getCriteria() {
    if(criteria.length() > 0)
        return criteria.substring(0,criteria.length()-5);
    else
        return "true";
    }
    public QueryCriteria() {
    }
    public QueryCriteria(Vector<String> columnNames) {
        this.columnNames=columnNames;
    }
    public void setColumnMatch(String column, String operator) {
        criteria.append(column);
        criteria.append(" ");
        criteria.append(operator);
        criteria.append(" ");
        criteria.append("?");
        criteria.append(" and ");
    }
    public void setColumnMatch(int columnIndex, String operator) {
        criteria.append(columnNames.get(columnIndex));
        criteria.append(" ");
        criteria.append(operator);
        criteria.append(" ");
        criteria.append("?");
        criteria.append(" and ");
    }
}
```

(3) 查询列 QueryColumn。

QueryColumn 用于定义查询面板中可供选择的数据库字段名及其标签名称。

```java
public class QueryColumn{
    private String lblName;
    private String dbColumn;
    public QueryColumn(String lblName, String dbColumn) {
        this.lblName=lblName;
        this.dbColumn=dbColumn;
    }
    public String toString() {
        return lblName;
    }
}
```

(4) 运算符 Operator。

Operator 用于定义查询界面中可供选择的运算符及其标签名称。

```java
public class Operator {
    private String name;
    private String value;
    public Operator(String name, String value) {
        this.name=name;
        this.value=value;
    }
    public String toString() {
        return name;
    }
}
```

6. 客户管理面板

客户管理面板 CustomManageGUI 用于实现客户的新建、修改与查询操作，效果如图 15-23 所示。

核心代码如下所示。

```java
public class CustomManageGUI extends RefreshPanel implements ActionListener{
    private InformationPal customInfo;
    private List<String> lblItemNames;
    private List<String> dbItemNames;
    private List<JComponent> cptItems;
    private List<JButton> btnCommond;
    private CustomService customServ;
    public CustomManageGUI() {
        lblItemNames=new ArrayList<String>();
        lblItemNames.add("身份证号码");
```

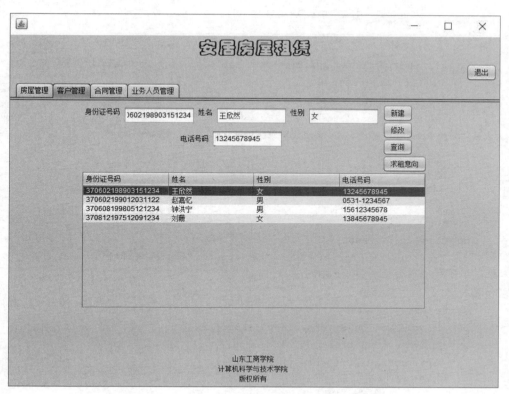

图 15-23　客户管理面板

```
lblItemNames.add("姓名");
lblItemNames.add("性别");
lblItemNames.add("电话号码");
dbItemNames=new ArrayList<String>();
dbItemNames.add("custom.personalId");
dbItemNames.add("custom.name");
dbItemNames.add("custom.gender");
dbItemNames.add("custom.telephone");
cptItems=new ArrayList<JComponent>();
for(int i=0; i<lblItemNames.size();i++) {
    JTextField t=new JTextField(10);
    t.setMaximumSize(t.getPreferredSize());
    cptItems.add(t);
}
btnCommond=new ArrayList<JButton>();
btnCommond.add(new JButton("新建"));
btnCommond.add(new JButton("修改"));
btnCommond.add(new JButton("查询"));
btnCommond.add(new JButton("求租意向"));
for(int i=0; i <btnCommond.size();i++) {
```

```java
            btnCommond.get(i).addActionListener(this);
        }
        try {
            customInfo=new InformationPal(lblItemNames,cptItems,btnCommond,3);
        } catch (Exception e) {
            e.printStackTrace();
        }
        customServ=new CustomServiceImpl(new CustomDaoImpl());
        refresh();
        customInfo.setDetailPanelSize(new Dimension(400,500));
        add(customInfo);
    }
    @Override
    public void refresh() {
        List<CustomInformation>customs=customServ.searchAllCustoms();
        customInfo.freshTable(customs);
    }
    @Override
    public void actionPerformed(ActionEvent e) {
        String strCmd=e.getActionCommand();
        JButton btn=(JButton)e.getSource();
        if("新建".equals(strCmd)){
            clear();
            ((JTextField)cptItems.get(0)).setEditable(true);
            btn.setText("保存");
        }else if("保存".equals(strCmd)) {
            ((JTextField)cptItems.get(0)).setEditable(false);
            CustomInformation custom=getInputCustomInfo();
            if(custom==null)
                return;
            btn.setText("新建");
            if(customServ.addCustom(custom)) {
                JOptionPane.showMessageDialog(CustomManageGUI.this, "创建成功");
                refresh();
            }else
                JOptionPane.showMessageDialog(CustomManageGUI.this, "创建失败");
        }else if("查询".equals(strCmd)) {
            Vector<QueryColumn>columnNames=new Vector<QueryColumn>();
            for(int i=0; i<dbItemNames.size();i++)
                columnNames.add(new QueryColumn(lblItemNames.get(i)
                                        , dbItemNames.get(i)));
            Vector<Operator>opers=new Vector<Operator>();
            for(int i=0; i<Constants.OPERATORS.length;i++)
                opers.add(Constants.OPERATORS[i]);
```

```java
        QueryCriteriaInputGUI query=new QueryCriteriaInputGUI(
                                        columnNames, opers);
        query.setBounds(CustomManageGUI.this.getX(), CustomManageGUI
                                        .this.getY(), 600, 400);
        query.setModal(true);
        query.setVisible(true);
        QueryCriteria criteria=query.getCriteria();
        List<Object>opt=query.getOpt();
        List<CustomInformation>customs=customServ.searchCustoms
                                        (criteria, opt);
        if(customs==null || customs.size()==0)
            JOptionPane.showMessageDialog(CustomManageGUI.this
                                        , "没有满足条件的记录");
        customInfo.freshTable(customs);
    }else if("修改".equals(strCmd)) {
        int confirm=JOptionPane.showConfirmDialog(CustomManageGUI.this
                                        , "确认修改当前承租人信息吗?");
        if(confirm==JOptionPane.YES_OPTION) {
            CustomInformation custom=getInputCustomInfo();
            if(custom==null)
                return;
            if(customServ.updateCustom(custom)) {
                JOptionPane.showMessageDialog(CustomManageGUI.this
                                        , "修改成功");
                refresh();
            }else
                JOptionPane.showMessageDialog(CustomManageGUI.this
                                        , "修改失败");
        }
    }else if("求租意向".equals(strCmd)) {
        String strId=((JTextField)cptItems.get(0)).getText();
        if(strId==null || strId.equals("")) {
            JOptionPane.showMessageDialog(CustomManageGUI.this
                                        ,"请选择当前客户");
            return;
        }
        JDialog intention=new JDialog();
        intention.setTitle("求租意向管理");
        IntentionManageGUI intentionPal=new IntentionManageGUI(strId);
        intention.add(intentionPal);
        intention.setBounds(CustomManageGUI.this.getX(),CustomManageGUI
                                        .this.getY(), 600, 400);
        intention.setModal(true);
        intention.setVisible(true);
```

```java
        }
    }
    public void clear() {
        for(int i=0; i<cptItems.size(); i++)
            ((JTextComponent)cptItems.get(i)).setText("");
    }
    private CustomInformation getInputCustomInfo() {
        String id=((JTextField)cptItems.get(0)).getText();
        String name=((JTextField)cptItems.get(1)).getText();
        String gender=((JTextField)cptItems.get(2)).getText();
        String telephone=((JTextField)cptItems.get(3)).getText();
        try {
            return new CustomInformation(id,name,gender,telephone);
        } catch (Exception ee) {
            JOptionPane.showMessageDialog(CustomManageGUI.this, ee.getMessage());
            return null;
        }
    }
}
```

7. 求租意向管理面板

求租意向管理面板 IntentionManageGUI 用于实现当前客户求租意向的创建、删除、修改操作，效果如图 15-24 所示。

核心代码如下所示。

```java
public class IntentionManageGUI extends RefreshPanel implements ActionListener{
    private InformationPal intentionInfo;
    private List<String> lblItemNames;
    private List<JComponent> cptItems;
    private List<JButton> btnCommond;
    private IntentionService intentionServ;
    private String tenantId;
    public IntentionManageGUI( String tenantId) {
        this.tenantId=tenantId;
        lblItemNames=new ArrayList<String>();
        lblItemNames.add("编号");
        lblItemNames.add("房型");
        lblItemNames.add("面积");
        lblItemNames.add("装修类型");
        lblItemNames.add("价格");
        lblItemNames.add("备注");
        cptItems=new ArrayList<JComponent>();
        for(int i=0; i<3;i++) {
            JTextField t=new JTextField(10);
```

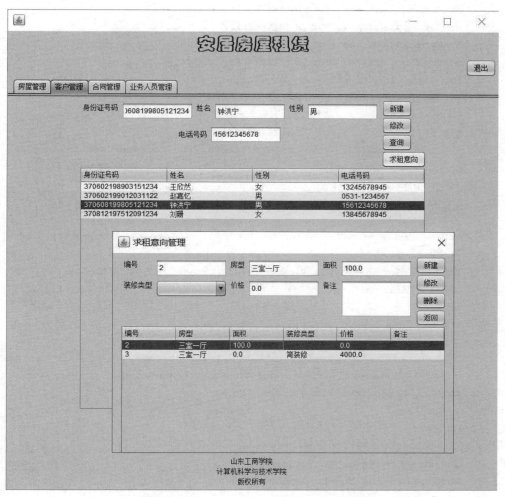

图 15-24 求租意向管理面板

```
    t.setMaximumSize(t.getPreferredSize());
    cptItems.add(t);
}
JComboBox com1=new JComboBox(Constants.STYLE_DECORATING);
com1.setPreferredSize(cptItems.get(0).getPreferredSize());
com1.setMaximumSize(com1.getPreferredSize());
cptItems.add(com1);
JTextField t=new JTextField(10);
t.setMaximumSize(t.getPreferredSize());
cptItems.add(t);
JTextArea remark=new JTextArea(3,10);
cptItems.add(remark);
((JTextField)cptItems.get(0)).setEditable(false);
btnCommond=new ArrayList<JButton>();
```

```java
            btnCommond.add(new JButton("新建"));
            btnCommond.add(new JButton("修改"));
            btnCommond.add(new JButton("删除"));
            btnCommond.add(new JButton("返回"));
            for(int i=0; i<btnCommond.size();i++) {
                btnCommond.get(i).addActionListener(this);
            }
            try {
                    intentionInfo=new InformationPal(lblItemNames,cptItems,
                                                    btnCommond,3);
            } catch (Exception e) {
                e.printStackTrace();
            }
            intentionServ=new IntentionServiceImpl(new IntentionDaoImpl());
            refresh();
            intentionInfo.setDetailPanelSize(new Dimension(500,500));
            add(intentionInfo);
        }
        @Override
        public void refresh() {
            List<IntentionInformation>intentions=intentionServ
                    .searchAllIntentionsByTenantId(tenantId);
            intentionInfo.freshTable(intentions);
        }
        @Override
        public void actionPerformed(ActionEvent e) {
            String strCmd=e.getActionCommand();
            JButton btn=(JButton)e.getSource();
            if("新建".equals(strCmd)){
                clear();
                btn.setText("保存");
            }else if("保存".equals(strCmd)) {
                IntentionInformation intention=getInputIntentionInfo();
                if(intention==null)
                    return;
                btn.setText("新建");
                if(intentionServ.addIntention(intention)) {
                    JOptionPane.showMessageDialog(IntentionManageGUI.this,"创建
                                                    成功");
                    refresh();
                }else
                    JOptionPane.showMessageDialog(IntentionManageGUI.this,"创建
                                                    失败");
                }else if("删除".equals(strCmd)) {
```

```java
            String strId=((JTextField)cptItems.get(0)).getText();
            if(strId==null || strId.equals("")) {
                JOptionPane.showMessageDialog(IntentionManageGUI.this
                                    ,"请选择待删除的意向");
            return;
            }
            int confirm=JOptionPane.showConfirmDialog(
                    IntentionManageGUI.this, "确认删除当前意向信息吗?");
            if(confirm==JOptionPane.YES_OPTION) {
                if(intentionServ.deleteIntentionById(strId)) {
                    JOptionPane.showMessageDialog(
                                IntentionManageGUI.this,"删除成功");
                    clear();
                    refresh();
                }else
                    JOptionPane.showMessageDialog(
                                IntentionManageGUI.this,"删除失败");
            }
        }else if("修改".equals(strCmd)) {
            int confirm=JOptionPane.showConfirmDialog(
                    IntentionManageGUI.this, "确认修改当前求租意向吗?");
            if(confirm==JOptionPane.YES_OPTION) {
                IntentionInformation intention=getInputIntentionInfo();
                if(intention==null)
                    return;
                if(intentionServ.updateIntention(intention)) {
                    JOptionPane.showMessageDialog(IntentionManageGUI.this
                                    , "修改成功");
                    refresh();
                }else
                    JOptionPane.showMessageDialog(IntentionManageGUI.this
                                    , "修改失败");
            }
        }else if("返回".equals(strCmd)){
            ((JDialog)(IntentionManageGUI.this.getTopLevelAncestor())).dispose();
        }
    }
    public void clear() {
        for(int i=0; i<3; i++)
            ((JTextComponent)cptItems.get(i)).setText("");
        ((JComboBox)cptItems.get(3)).setSelectedItem("");
        for(int i=4; i<cptItems.size(); i++)
            ((JTextComponent)cptItems.get(i)).setText("");
```

```java
    }
    private IntentionInformation getInputIntentionInfo() {
        int id=0;
        String strId=((JTextField)cptItems.get(0)).getText();
        if(!strId.equals(""))
            id=Integer.parseInt(strId);
        String layout=((JTextField)cptItems.get(1)).getText();
        float area=0;
        String strArea=((JTextField)cptItems.get(2)).getText();
        if(!strArea.equals(""))
            area=Float.parseFloat(strArea);
        String decoration=(String)(((JComboBox)cptItems.get(3))
                                        .getSelectedItem());
        float price=0;
        String strPrice=((JTextField)cptItems.get(4)).getText();
        if(!strPrice.equals(""))
            price=Float.parseFloat(strPrice);
        String remark=(String)(((JTextArea)cptItems.get(5)).getText());
        try {
            return new IntentionInformation(id,tenantId,layout,area,
            decoration,price,remark);
        } catch (Exception ee) {
            JOptionPane.showMessageDialog(IntentionManageGUI.this,
            ee.getMessage());
                return null;
        }
    }
}
```

8. 合同管理面板

合同管理面板 ContractManageGUI 用于实现签订合同、提前解约及查询合同操作，效果图如图 15-25 所示。

核心代码如下所示。

```java
public class ContractManageGUI extends RefreshPanel implements ActionListener {
    private ContractService contractServ;
    private HouseService houseServ;
    private CustomService customServ;
    private SalesmanService salesmanServ;
    private InformationPal contractInfo;
    private List<String> lblItemNames;
    private List<String> dbItemNames;
    private List<JComponent> cptItems;
    private List<JButton> btnCommond;
```

第15章 房屋租赁系统的设计与实现

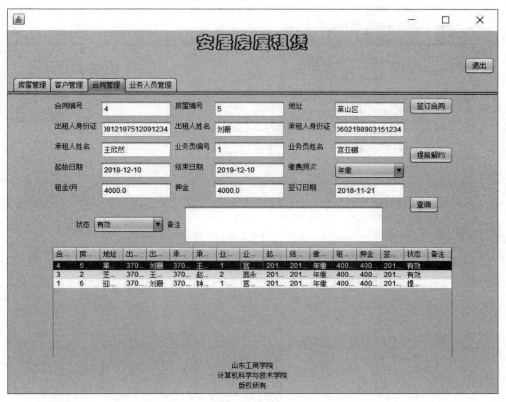

图 15-25 合同管理面板

```
private HouseInformation house;
private CustomInformation tenant;
private SalesmanInformation salesman;
public ContractManageGUI() {
    lblItemNames=new ArrayList<String>();
    lblItemNames.add("合同编号");
    lblItemNames.add("房屋编号");
    lblItemNames.add("地址");
    lblItemNames.add("出租人身份证");
    lblItemNames.add("出租人姓名");
    lblItemNames.add("承租人身份证");
    lblItemNames.add("承租人姓名");
    lblItemNames.add("业务员编号");
    lblItemNames.add("业务员姓名");
    lblItemNames.add("起始日期");
    lblItemNames.add("结束日期");
    lblItemNames.add("缴费频次");
    lblItemNames.add("租金/月");
    lblItemNames.add("押金");
```

```java
lblItemNames.add("签订日期");
lblItemNames.add("状态");
lblItemNames.add("备注");
dbItemNames=new ArrayList<String>();
dbItemNames.add("contract.id");
dbItemNames.add("contract.houseId");
dbItemNames.add("house.address");
dbItemNames.add("custom.personalId");
dbItemNames.add("custom.name");
dbItemNames.add("custom.personalId");
dbItemNames.add("custom.name");
dbItemNames.add("salesman.id");
dbItemNames.add("salesman.name");
dbItemNames.add("contract.startDate");
dbItemNames.add("contract.endDate");
dbItemNames.add("contract.freOfPay");
dbItemNames.add("contract.rentOfMonth");
dbItemNames.add("contract.deposit");
dbItemNames.add("contract.signDate");
dbItemNames.add("contract.state");
dbItemNames.add("contract.remark");
cptItems=new ArrayList<JComponent>();
for (int i=0; i <11; i++) {
    JTextField t=new JTextField(10);
    t.setMaximumSize(t.getPreferredSize());
    cptItems.add(t);
}
((JTextField) cptItems.get(0)).setEditable(false);
JComboBox com1=new JComboBox(Constants.FREQUENCY_PAY);
com1.setPreferredSize(cptItems.get(0).getPreferredSize());
com1.setMaximumSize(com1.getPreferredSize());
cptItems.add(com1);
for (int i=12; i <15; i++) {
    JTextField t=new JTextField(10);
    t.setMaximumSize(t.getPreferredSize());
    cptItems.add(t);
}
JComboBox com2=new JComboBox(Constants.STATE_CONTRACT);
com2.setPreferredSize(cptItems.get(0).getPreferredSize());
com2.setMaximumSize(com2.getPreferredSize());
cptItems.add(com2);
JTextArea remark=new JTextArea(3, 30);
cptItems.add(remark);
JTextField tfdHouseID=(JTextField) cptItems.get(1);
```

```java
tfdHouseID.addActionListener(new ActionListener() {
    @Override
    public void actionPerformed(ActionEvent e) {
        //根据房屋编号查询房屋
        house=houseServ.searchHouseById(tfdHouseID.getText());
        if (house==null) {
            JOptionPane.showMessageDialog(ContractManageGUI.this
                                        , "房屋编号不正确");
            return;
        }
        ((JTextField) cptItems.get(2)).setText(house.getAddress());
        ((JTextField) cptItems.get(3)).setText(house.getOwnerId());
        ((JTextField) cptItems.get(4)).setText(house.getOwnerName());
    }
});
JTextField tfdTenantID=(JTextField) cptItems.get(5);
tfdTenantID.addActionListener(new ActionListener() {
    @Override
    public void actionPerformed(ActionEvent e) {
        //根据承租人身份证号码查询承租人
        tenant=customServ.searchCustomByPersonalId(tfdTenantID.getText());
        if (tenant==null) {
            JOptionPane.showMessageDialog(ContractManageGUI.this
                                        , "承租人身份证号码不正确");
            return;
        }
        ((JTextField) cptItems.get(6)).setText(tenant.getName());
    }
});
JTextField tfdSalesmanID=(JTextField) cptItems.get(7);
tfdSalesmanID.addActionListener(new ActionListener() {
    @Override
    public void actionPerformed(ActionEvent e) {
        //根据业务员工号查询业务员
        salesman=salesmanServ.searchSalesmanById(tfdSalesmanID.getText());
        if (salesman==null) {
            JOptionPane.showMessageDialog(ContractManageGUI.this
                                        , "业务员工号不正确");
            return;
        }
        ((JTextField) cptItems.get(8)).setText(salesman.getName());
    }
});
btnCommond=new ArrayList<JButton>();
```

```java
        btnCommond.add(new JButton("签订合同"));
        btnCommond.add(new JButton("提前解约"));
        btnCommond.add(new JButton("查询"));
        for (int i=0; i <btnCommond.size(); i++) {
            btnCommond.get(i).addActionListener(this);
        }
        try {
            contractInfo=new InformationPal(lblItemNames, cptItems,
                btnCommond, 3);
        } catch (Exception e) {
            e.printStackTrace();
        }
        houseServ=new HouseServiceImpl(new HouseDaoImpl());
        customServ=new CustomServiceImpl(new CustomDaoImpl());
        salesmanServ=new SalesmanServiceImpl(new SalesmanDaoImpl());
        contractServ=new ContractServiceImpl(new ContractDaoImpl(),
            houseServ);
        refresh();
        contractInfo.setDetailPanelSize(new Dimension(800, 500));
        add(contractInfo);
    }
    public void refresh() {
        List<ContractInformation>contracts=contractServ.searchAllContracts();
        contractInfo.freshTable(contracts);
    }
    @Override
    public void actionPerformed(ActionEvent e) {
        String strCmd=e.getActionCommand();
        JButton btn= (JButton) e.getSource();
        if ("签订合同".equals(strCmd)) {
            clear();
            btn.setText("保存");
        } else if ("保存".equals(strCmd)) {
            ContractInformation contract=getInputContractInfo();
            if (contract==null)
                return;
            btn.setText("签订合同");
            try {
                ContractDB c=new ContractDB(contract);
                if (contractServ.addContract(c)) {
                    JOptionPane.showMessageDialog(ContractManageGUI.this
                                            , "合同创建成功");
                    refresh();
                    ((JTextField) cptItems.get(0)).setText("" +c.getId());
```

```java
            } else
                JOptionPane.showMessageDialog(ContractManageGUI.this
                                            , "合同创建失败");
        } catch (Exception e1) {
            JOptionPane.showMessageDialog(ContractManageGUI.this
                                        , e1.getMessage());
        }
    } else if ("查询".equals(strCmd)) {
        Vector<QueryColumn>columnNames=new Vector<QueryColumn>();
        for (int i=0; i <dbItemNames.size(); i++)
            columnNames.add(new QueryColumn(lblItemNames.get(i)
                                    , dbItemNames.get(i)));
        Vector<Operator>opers=new Vector<Operator>();
        for (int i=0; i <Constants.OPERATORS.length; i++)
            opers.add(Constants.OPERATORS[i]);
        QueryCriteriaInputGUI query=new QueryCriteriaInputGUI(
                                    columnNames, opers);
        query.setBounds(ContractManageGUI.this.getX(), ContractManageGUI
                                    .this.getY(), 600, 400);
        query.setModal(true);
        query.setVisible(true);
        QueryCriteria criteria=query.getCriteria();
        List<Object>opt=query.getOpt();
        List<ContractInformation>contracts=contractServ.searchContracts
                                    (criteria, opt);
        if (contracts==null || contracts.size()==0)
            JOptionPane.showMessageDialog(ContractManageGUI.this
                                    , "没有满足条件的记录");
        contractInfo.freshTable(contracts);
    } else if ("提前解约".equals(strCmd)) {
        ContractInformation contract=getInputContractInfo();
        if (contract==null)
            return;
        int confirm=
            JOptionPane.showConfirmDialog(ContractManageGUI.this
                                    , "确认解约当前合同吗?");
        if (confirm==JOptionPane.YES_OPTION) {
            if (contractServ.teminateContractById(contract)) {
                JOptionPane.showMessageDialog(
                            ContractManageGUI.this, "解约成功");
                //刷新界面
                ((JComboBox) cptItems.get(15)).setSelectedItem(
                            Constants.STATE_CONTRACT[2]);
                refresh();
```

```java
                } else
                    JOptionPane.showMessageDialog(
                            ContractManageGUI.this, "无法解约");
            }
        }
    }
    public void clear() {
        for (int i=0; i <11; i++)
            ((JTextField) cptItems.get(i)).setText("");
        ((JComboBox) cptItems.get(11)).setSelectedItem("");
        for (int i=12; i <15; i++)
            ((JTextField) cptItems.get(i)).setText("");
        ((JComboBox) cptItems.get(15)).setSelectedItem("");
        ((JTextArea) cptItems.get(16)).setText("");
    }
    private ContractInformation getInputContractInfo() {
        String strId=((JTextField) cptItems.get(0)).getText();
        String strHouseId=((JTextField) cptItems.get(1)).getText();
        String address=((JTextField) cptItems.get(2)).getText();
        String tenantId=((JTextField) cptItems.get(5)).getText();
        String tenantName=((JTextField) cptItems.get(6)).getText();
        String ownerId=((JTextField) cptItems.get(3)).getText();
        String ownerName=((JTextField) cptItems.get(4)).getText();
        String strSalesmanId=((JTextField) cptItems.get(7)).getText();
        String salesmanName=((JTextField) cptItems.get(8)).getText();
        String startDate=((JTextField) cptItems.get(9)).getText();
        String endDate=((JTextField) cptItems.get(10)).getText();
        String freOfPay=(String) (((JComboBox) cptItems.get(11)).
        getSelectedItem());
        String strRentOfMonth=((JTextField) cptItems.get(12)).getText();
        String strDepoist=((JTextField) cptItems.get(13)).getText();
        String signDate=((JTextField) cptItems.get(14)).getText();
        String state=(String) (((JComboBox) cptItems.get(15))
                                    .getSelectedItem());
        String remark=((JTextArea) cptItems.get(16)).getText();
        if (house==null)
            house=houseServ.searchHouseById(strHouseId);
        if (house==null || !house.getAddress().equals(address)) {
            JOptionPane.showMessageDialog(ContractManageGUI.this
                                , "房屋信息不正确,请确认");
            return null;
        }
        if (!house.getOwnerId().equals(ownerId)
                || !house.getOwnerName().equals(ownerName)) {
```

```java
        JOptionPane.showMessageDialog(ContractManageGUI.this
                                    ,"出租人信息不正确,请确认");
        return null;
    }
    if (tenant==null)
        tenant=customServ.searchCustomByPersonalId(tenantId);
    if (tenant==null || !tenant.getName().equals(tenantName)) {
        JOptionPane.showMessageDialog(ContractManageGUI.this
                                    ,"承租人信息不正确,请确认");
        return null;
    }
    if (salesman==null)
        salesman=salesmanServ.searchSalesmanById(strSalesmanId);
    if (salesman==null || !salesman.getName().equals(salesmanName)) {
        JOptionPane.showMessageDialog(ContractManageGUI.this
                                    ,"业务员信息不正确,请确认");
        return null;
    }
    house=null;
    tenant=null;
    salesman=null;
    if (startDate==null || startDate.equals("") || endDate==null
        || endDate.equals("") || freOfPay==null || freOfPay.equals("")
        || strRentOfMonth==null || strRentOfMonth.equals("")
        || strDepoist==null || strDepoist.equals("") || signDate==null
        || signDate.equals("") || state==null || state.equals("")) {
        JOptionPane.showMessageDialog(ContractManageGUI.this
                                    ,"合同信息不完整,请确认");
        return null;
    }
    int id=0;
    if (!strId.equals(""))
        id=Integer.parseInt(strId);
    int houseId=Integer.parseInt(strHouseId);
    int salemanId=Integer.parseInt(strSalesmanId);
    float rentOfMonth=Float.parseFloat(strRentOfMonth);
    float deposit=Float.parseFloat(strDepoist);
    try {
        return new ContractInformation(id, houseId, address, ownerId
                , ownerName, tenantId, tenantName, salemanId
                ,salesmanName, startDate, endDate, freOfPay, rentOfMonth
                , deposit, signDate, state, remark);
    } catch (Exception e) {
```

```
                    JOptionPane.showMessageDialog(ContractManageGUI.this,
                                                            e.getMessage());
                    return null;
                }
            }
        }
```

9. 业务人员管理面板

业务人员管理面板 SalesmanManageGUI 用于实现业务人员的入职、离职、修改及查询操作,效果如图 15-26 所示。

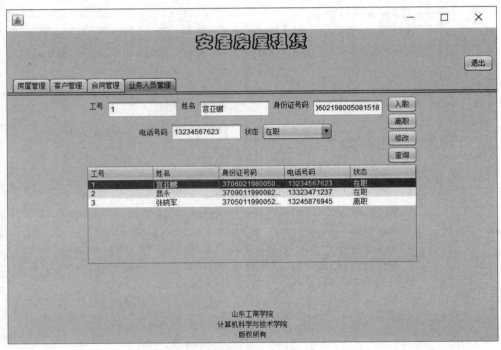

图 15-26 业务人员管理面板

核心代码如下所示。

```
public class SalesmanManageGUI extends RefreshPanel implements ActionListener {
    private SalesmanService salesmanServ;
    private InformationPal salesmanInfo;
    private List<String>lblItemNames;
    private List<String>dbItemNames;
    private List<JComponent>cptItems;
    private List<JButton>btnCommond;
    public SalesmanManageGUI() {
        lblItemNames=new ArrayList<String>();
```

```java
        lblItemNames.add("工号");
        lblItemNames.add("姓名");
        lblItemNames.add("身份证号码");
        lblItemNames.add("电话号码");
        lblItemNames.add("状态");
        dbItemNames=new ArrayList<String>();
        dbItemNames.add("salesman.id");
        dbItemNames.add("salesman.name");
        dbItemNames.add("salesman.personalId");
        dbItemNames.add("salesman.telephone");
        dbItemNames.add("salesman.state");
        cptItems=new ArrayList<JComponent>();
        for (int i=0; i<4; i++) {
            JTextField t=new JTextField(10);
            t.setMaximumSize(t.getPreferredSize());
            cptItems.add(t);
        }
        JComboBox com1=new JComboBox(Constants.STATE_SALESMAN);
        com1.setPreferredSize(cptItems.get(0).getPreferredSize());
        com1.setMaximumSize(com1.getPreferredSize());
        cptItems.add(com1);
        ((JTextField) cptItems.get(0)).setEditable(false);
        btnCommond=new ArrayList<JButton>();
        btnCommond.add(new JButton("入职"));
        btnCommond.add(new JButton("离职"));
        btnCommond.add(new JButton("修改"));
        btnCommond.add(new JButton("查询"));
        for (int i=0; i<btnCommond.size(); i++) {
            btnCommond.get(i).addActionListener(this);
        }
        try {
            salesmanInfo=new InformationPal(lblItemNames, cptItems,
            btnCommond, 3);
        } catch (Exception e) {
            e.printStackTrace();
        }
        salesmanServ=new SalesmanServiceImpl(new SalesmanDaoImpl());
        refresh();
        salesmanInfo.setDetailPanelSize(new Dimension(400, 500));
        add(salesmanInfo);
    }
    public void refresh() {
        List<SalesmanInformation> owners=salesmanServ.searchAllSalesman();
        salesmanInfo.freshTable(owners);
```

```java
        }
        @Override
        public void actionPerformed(ActionEvent e) {
            String strCmd=e.getActionCommand();
            JButton btn=(JButton) e.getSource();
            if ("入职".equals(strCmd)) {
                clear();
                btn.setText("保存");
            } else if ("离职".equals(strCmd)) {
                int confirm=JOptionPane.showConfirmDialog(
                        SalesmanManageGUI.this,"确认当前业务人员离职?");
                if (confirm==JOptionPane.YES_OPTION) {
                    //修改状态为"已删除"
                    String id=((JTextField) cptItems.get(0)).getText();
                    if (id==null || id.equals("")) {
                        JOptionPane.showMessageDialog(SalesmanManageGUI.this
                                ,"请选择待离职员工");
                        return;
                    }
                    if (salesmanServ.deleteSalesmanById(id)) {
                        JOptionPane.showMessageDialog(SalesmanManageGUI.this
                                ,"离职成功");
                        clear();
                        refresh();
                    } else
                        JOptionPane.showMessageDialog(SalesmanManageGUI.this
                                ,"离职失败");
                }
            } else if ("保存".equals(strCmd)) {
                SalesmanInformation salesman=getInputInfo();
                if (salesman==null)
                    return;
                btn.setText("入职");
                if (salesmanServ.addSalesman(salesman)) {
                    JOptionPane.showMessageDialog(SalesmanManageGUI.this,
                            "创建成功");
                    refresh();
                    ((JTextField) cptItems.get(0)).setText("" +salesman.getId());
                } else
                    JOptionPane.showMessageDialog(SalesmanManageGUI.this,
                            "创建失败");
            } else if ("查询".equals(strCmd)) {
                Vector<QueryColumn>columnNames=new Vector<QueryColumn>();
                for (int i=0; i<dbItemNames.size(); i++)
```

```java
            columnNames.add(new QueryColumn(lblItemNames.get(i)
                                        , dbItemNames.get(i)));
        Vector<Operator>opers=new Vector<Operator>();
        for (int i=0; i<Constants.OPERATORS.length; i++)
            opers.add(Constants.OPERATORS[i]);
        QueryCriteriaInputGUI query=new QueryCriteriaInputGUI(
                                        columnNames, opers);
        query.setBounds(SalesmanManageGUI.this.getX()
                        , SalesmanManageGUI.this.getY(), 600, 400);
        query.setModal(true);
        query.setVisible(true);
        QueryCriteria criteria=query.getCriteria();
        List<Object>opt=query.getOpt();
        List<SalesmanInformation>salesman=salesmanServ.searchSalesman(
                                        criteria, opt);
        if (salesman==null || salesman.size()==0)
            JOptionPane.showMessageDialog(SalesmanManageGUI.this
                                        , "没有满足条件的记录");
        salesmanInfo.freshTable(salesman);
    } else if ("修改".equals(strCmd)) {
        int confirm=JOptionPane.showConfirmDialog(
                SalesmanManageGUI.this, "确认修改当前业务人员信息吗?");
        if (confirm==JOptionPane.YES_OPTION) {
            SalesmanInformation salesman=getInputInfo();
            if (salesman==null)
                return;
            if (salesmanServ.updateSalesman(salesman)) {
                JOptionPane.showMessageDialog(SalesmanManageGUI.this
                                        , "修改成功");
                refresh();
            } else
                JOptionPane.showMessageDialog(SalesmanManageGUI.this
                                        , "修改失败");
        }
    }
}
public void clear() {
    for (int i=0; i<4; i++)
        ((JTextField) cptItems.get(i)).setText("");
    ((JComboBox) cptItems.get(4)).setSelectedItem("");
}
private SalesmanInformation getInputInfo() {
    int id=0;
    String strId=((JTextField) cptItems.get(0)).getText();
```

```
        if (!strId.equals(""))
            id=Integer.parseInt(strId);
        String name=((JTextField) cptItems.get(1)).getText();
        String personalId=((JTextField) cptItems.get(2)).getText();
        String telephone=((JTextField) cptItems.get(3)).getText();
        String state=(String) ((JComboBox) cptItems.get(4)).getSelectedItem();
        try {
            return new SalesmanInformation(id, name, personalId, telephone,
                state);
        } catch (Exception e) {
            JOptionPane.showMessageDialog(SalesmanManageGUI.this,
                e.getMessage());
            return null;
        }
    }
}
```

15.4 测试

利用 JUnit 对业务服务包 service 包含的实现类逐一进行单元测试。

15.4.1 搭建测试环境——导入 JUnit 包

在 Eclipse 的 Package Explorer 窗口中选择待测试的项目，单击右键，依次选择 Build Path/Add Libraries 选项，弹出如图 15-27 所示的对话框。

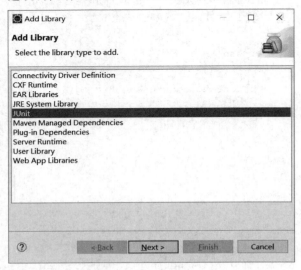

图 15-27　导入 JUnit 包

选择 JUnit，单击 Netx 按钮即可。

15.4.2 单元测试

以 HouseServeImpl 类为例，完成下列测试。

```java
package service.test;
class HouseServTest {
    private static HouseService houseServ;
    @BeforeAll
    static void setUpBeforeClass() throws Exception {
        houseServ=new HouseServiceImpl(new HouseDaoImpl());
    }
    @Test
    void testAddHouse() {
        for(int i=0; i<3;i++) {
            String ownerId="370602198903151234";
            HouseDB house=new HouseDB();
            house.setOwnerId(ownerId);
            house.setAddress("崂山区"+i+1);
            house.setLayout("一室一厅");
            house.setArea(100);
            house.setDecoration("精装修");
            house.setPrice(4000);
            house.setState("待出租");
            assertTrue(houseServ.addHouse(house));
        }
    }
    @Test
    void testSearchAllHouses() {
        List<HouseInformation>houses=houseServ.searchAllHouses();
        System.out.println(houses);
    }
    @Test
    void testUpdateHouse() {
        HouseDB h=new HouseDB(7,"370602198903151234","青岛市崂山区"
                    ,"一室一厅",100,"精装修",4000,"待出租","整租");
        assertTrue(houseServ.updateHouse(h));
    }
    @Test
    void testDeleteHouseById() {
        assertTrue(houseServ.deleteHouseById("8"));
    }
    @Test
    void testSearchHouses() {
```

```java
        QueryCriteria criteria=new QueryCriteria();
        criteria.setColumnMatch("house.ownerId", "=");
        List<Object> opt=new ArrayList<Object>();
        opt.add("370602198903151234");
        List<HouseInformation> houses=houseServ.searchHouses(criteria, opt);
        System.out.println(houses);
    }
    @Test
    void testSearchHouseById() {
        HouseInformation house=houseServ.searchHouseById("7");
        System.out.println(house);
    }
    @Test
    void testUpdateHouseStateById() {
        assertTrue(houseServ.updateHouseStateById(7, "已出租"));
    }
}
```

15.5 程序发布

15.5.1 打包项目

1. 创建清单文件 Mainifest.mf

```
Manifest-Version: 1.0
Class-Path: lib/mysql-connector-java-8.0.20.jar
Main-Class: userInterface.LoginGUI
```

清单文件的书写规范如下：
① 第一行不能空，行与行之间不能有空行；
② 每一行的最后一个字符不能是空格；
③ 最后一行一定是空行；
④ 每个属性的名称和值之间（即冒号后面）一定要有空格；
⑤ 文件的每一行都不能超过 72 字节。

2. 在 Eclipse 中对项目进行打包输出

选择待打包的项目，单击右键，选择 Export 选项，显示如图 15-28 所示的对话框。
单击 Next 按钮，显示如图 15-29 所示的对话框。
单击 Next 按钮，显示如图 15-30 所示的对话框。
单击 Next 按钮，显示如图 15-31 所示。
单击 Finsh 按钮，在 D 盘的根目录下生成了 HouseLeasing.jar 文件。

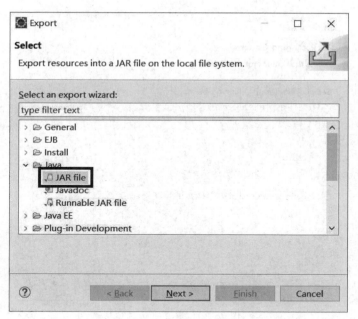

图 15-28　选择导出类型

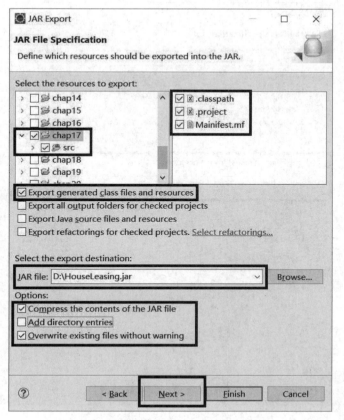

图 15-29　定义包文件

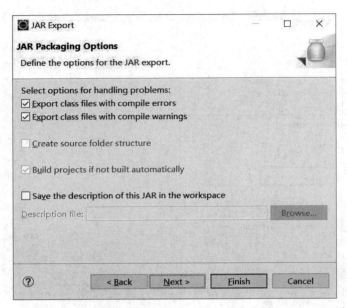

图 15-30 定义打包选项

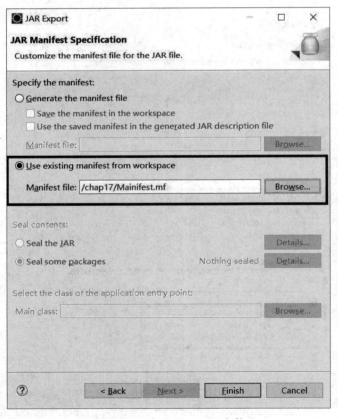

图 15-31 定义 Manifest 文件

15.5.2 部署项目

在 HouseLeasing.jar 所在的目录下创建 lib 文件夹，将 MySQL 的 Connector J 文件 mysql-connector-java-8.0.20.jar 复制到 lib 文件夹。

15.5.3 运行项目

在命令行状态下输入如图 15-32 所示的命令。

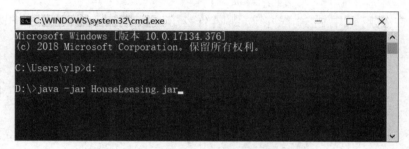

图 15-32　通过控制台运行房屋租赁系统软件

图书资源支持

感谢您一直以来对清华版图书的支持和爱护。为了配合本书的使用,本书提供配套的资源,有需求的读者请扫描下方的"书圈"微信公众号二维码,在图书专区下载,也可以拨打电话或发送电子邮件咨询。

如果您在使用本书的过程中遇到了什么问题,或者有相关图书出版计划,也请您发邮件告诉我们,以便我们更好地为您服务。

我们的联系方式:

地　　址:北京市海淀区双清路学研大厦 A 座 714

邮　　编:100084

电　　话:010-83470236　010-83470237

客服邮箱:2301891038@qq.com

QQ:2301891038(请写明您的单位和姓名)

资源下载:关注公众号"书圈"下载配套资源。

书圈

获取最新书目

观看课程直播